SpringerBriefs in Electrical and Computer Engineering

SpringerBriefs in Speech Technology

Studies in Speech Signal Processing, Natural Language
Understanding, and Machine Learning

Series Editor
Amy Neustein
Fort Lee, New Jersey, USA

SpringerBriefs present concise summaries of cutting-edge research and practical applications across a wide spectrum of fields. Featuring compact volumes of 50 to 125 pages, the series covers a range of content from professional to academic. Typical topics might include: timely report of state-of-the art analytical techniques, a bridge between new research results, as published in journal articles, and a contextual literature review, a snapshot of a hot or emerging topic, an in-depth case study or clinical example and a presentation of core concepts that students must understand in order to make independent contributions.

More information about this series at http://www.springer.com/series/10059

Koteswara Rao Anne • Swarna Kuchibhotla
Hima Deepthi Vankayalapati

Acoustic Modeling for Emotion Recognition

 Springer

Koteswara Rao Anne
Dept of Information Technology
Velagapudi Ramakrishna Siddhartha
  Engineering College
Vijayawada
India

Hima Deepthi Vankayalapati
Dept of Computer Science and Engineering
Velagapudi Ramakrishna Siddhartha
  Engineering College, Kanuru
Vijayawada
India

Swarna Kuchibhotla
Acharya Nagarjuna University
Andhra Pradesh
India

ISSN 2191-8112   ISSN 2191-8120 (electronic)
SpringerBriefs in Electrical and Computer Engineering
ISSN 2191-737X   ISSN 2191-7388 (electronic)
SpringerBriefs in Speech Technology
ISBN 978-3-319-15529-6   ISBN 978-3-319-15530-2 (eBook)
DOI 10.1007/978-3-319-15530-2

Library of Congress Control Number: 2015934453

Springer Cham Heidelberg New York Dordrecht London

Printed on acid-free paper

Springer is part of Springer Science+Business Media (www.springer.com)

Contents

Chapter 1
Introduction

1.1 Speech Signal Representation

The main aim of speech is for communication. The way of representing the speech is through a signal or waveform, which carries the message content or information. The speech signal is transmitted, stored, and processed in many ways in communication systems. First the message content of the speech signal should be kept safe and the next is representation of the speech signal in a convenient form which is more useful for transformation with out modifying or loss of message content in the speech signal. The speech produced by a human is analog signal which is continuously varying with time and is represented by $x(t)$. Here t represents time variable. For digital speech processing we can convert that analog speech signal into digital. In digital notation it can be represented as a sequence of samples $x(n)$. The more convenient form of representing the speech signal is by sum of sinusoids as it leads to convenient solution to problems like formant estimation, pitch period estimation etc., $x(n)$ can be represented by the Eq. 1.1

$$x(n) = cos(\omega_0 n) \tag{1.1}$$

where ω_0 is the frequency of the sinusoid. The Fourier representations provide convenient means to determine response to a sum of sinusoids for linear systems. The Discrete Time Fourier Transform and inverse fourier transform are given by the Eqs. 1.2 and 1.3.

$$X(e^{j\omega}) = \sum_{n=-\infty}^{\infty} x(n)e^{-j\omega n} \tag{1.2}$$

$$x(n) = \frac{1}{2\pi} \int_{-\pi}^{\pi} X(e^{j\omega})e^{j\omega n} \, dw \tag{1.3}$$

here $X(e^{j\omega})$ is continuous and periodic function with period 2π and ω is the frequency variable of $X(e^{j\omega})$. The sufficient condition for convergence is

© The Author(s) - SpringerBriefs 2015
K. R. Anne et al., *Acoustic Modeling for Emotion Recognition*,
SpringerBriefs in Electrical and Computer Engineering, DOI 10.1007/978-3-319-15530-2_1

$$\sum_{n=-\infty}^{\infty} |x(n)| < \infty \qquad (1.4)$$

The convolution expression can be represented as

$$y(n) = x(n) * h(n) \qquad (1.5)$$

$$Y(e^{jw}) = X(e^{jw})H(e^{jw}) \qquad (1.6)$$

As speech is not a stationary signal, the properties are changes with time. So it needs more sophisticated analysis to reflect time varying properties. For this we split the speech signal into a set of frames over a fixed time intervals like 10–30 msec approximately. The time dependent fourier transform or short time fourier transform of the speech signal is given by the Eq. 1.7

$$X_n(e^{j\omega}) = \sum_{m=-\infty}^{\infty} w(n-m)x(m)e^{-j\omega m} \qquad (1.7)$$

where $W(n-m)$ is a window function. The main aim of windowing is to emphasize a finite portion of a speech signal in the vicinity of sample n and de-emphasize the remainder portion of the speech signal. Here n is a discrete time index variable and ω is a frequency variable. If n is fixed then it can be shown that

$$X_n(e^{j\omega}) = \frac{1}{2\pi} \int_{-\pi}^{\pi} W(e^{j\theta})e^{j\theta n} X(e^{j(\omega+\theta)})d\theta \qquad (1.8)$$

The Eq. 1.8 is meaningful if we assume that $X(e^{j\omega})$ represents Fourier transform of a signal whose basic properties continue outside the window or simply the signal is zero outside the window.

With this the properties of window Fourier transform $W(e^{j\theta})$ become important. In order for faithful reproduction of properties of $X(e^{j\omega})$ in $X_n(e^{j\omega})$, the function $W(e^{j\theta})$ must resemble an impulse with respect to $X(e^{j\omega})$.

1.2   Acoustic Signal Basics

The way the speech signals are produced and perceived by the human is starting point of the research. Human speech communication system produces ideas (word sequence) which are made within the speaker brain. These word sequences are de-livered by his/her text generator. The general human vocal system is modeled by the speech generator. The speech generator converts the word sequence into speech signal and is transferred to listener through air or in other words speech is produced by air-pressure waves which are emanating from the mouth and nostrils of a speaker. At the listener side, the human auditory system receives these acoustic signals and listeners brain starts the processing of signal to understand its content. The speech

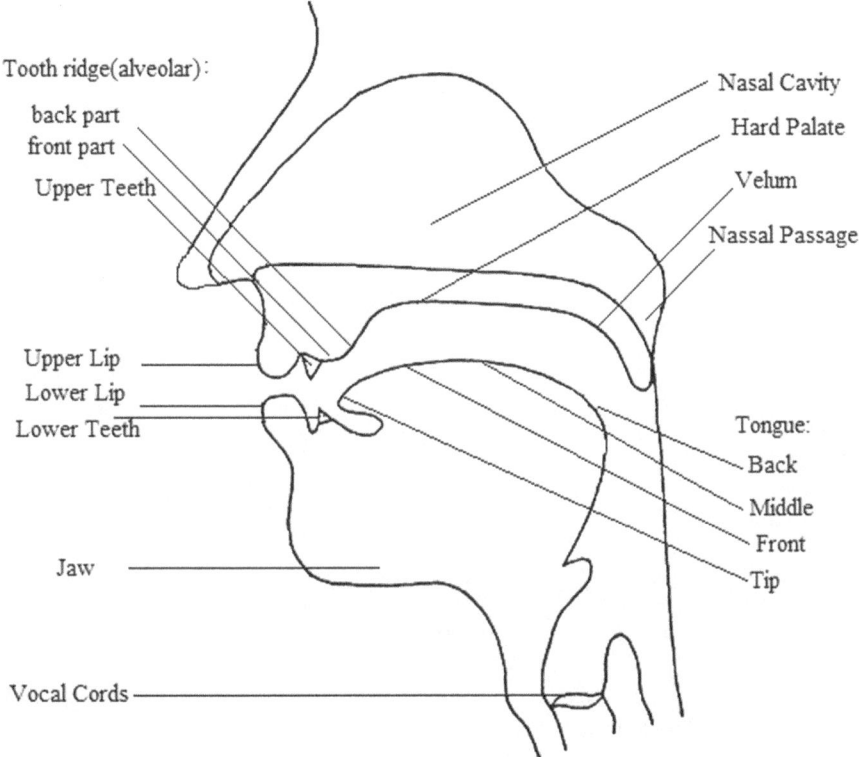

Tooth ridge(alveolar):

    back part
    front part

    Upper Teeth

Upper Lip
Lower Lip
Lower Teeth

Jaw

Vocal Cords

Nasal Cavity

Hard Palate

Velum

Nassal Passage

Tongue:

    Back

    Middle

    Front

    Tip

Fig. 1.1 Block diagram of human speech production system

recognizer modeled by the speech decoder, it decodes the acoustic signal into word sequence. So speech production and speech perception are in inverse processes in the speech recognition application. The human basic speech production system is shown in Fig. 1.1.

The components of speech production system are lungs, trachea, larynx, pharynx (throat),oral and nasal cavity. The pharynx and oral cavities are typically referred as vocal tract, and the nasal cavity is as the nasal tract. Lungs are the main source of air during speech. The Speech sounds are mainly classified into 2 types voiced and unvoiced. Sounds are said to be voiced when vocal folds come close together and oscillate against one another during speech sound. sounds are said to be unvoiced when the vocal folds are too slack to vibrate periodically. The place where the vocal folds come close together is called glottis. Velum is like a valve, opening to allow passage of air through the nasal cavity. A long and hard surface at the roof inside the mouth is called hard palate. It enables consonant articulation when tongue is placed against to it. Tongue is a flexible articulator, shaped away from the palate for vowels and placed close to the palate for consonant articulation. Another place of articulation which is used to brace the tongue for certain consonants is teeth. Lips can spread to affect vowel quality and closed completely to stop the oral air flow for certain consonants like p, m etc.,

Fig. 1.2 Block diagram for valence arousal space

Speech signal transmits mainly two types of messages. First, it contains the explicit message or information content about something or nothing. Second it contains the implicit message about the speaker like age, gender, emotion and health state etc. It is some how easy in extracting and understanding the content in the message with some technological advancements. However the task of understanding the human emotion in implicit message is a difficult task. Various signal processing tasks have been developed to tackle this task. Now our aim is to discuss those signal processing techniques and able to identify the correct human emotional state by the computer. The following section will discuss on this issue more.

1.3   Different Perspectives of Emotion

Construction of an emotion recognizer mainly depends upon sense of what actually an emotion is. Understanding the nature of emotion matters more in the context of automatic emotion recognition. This is done by extracting certain features which are relevant in describing the emotional state and distinguishing it from other emotional states. According to Darwin [14], emotions are distinctive action patterns selected by evolution because of their survival value.

Valence Arousal space is a representation of capturing a wide range of significant issues in emotion and is shown in the Fig. 1.2. The horizontal axis represents valance state in which the emotions of a person are influenced by positive or negative evaluations of people or things or events. The vertical axis represents the aroused state in which the activation level is described i.e., the strength of the human disposition to take some action rather than none.

1.3.1   Physiology of Emotion

Each emotion induces physiological changes which affect speech directly and attempts have been made to infer vocal signs of emotion on that basis [59]. Physiological changes include affects in measures of speech features like intensity, pitch and speech rate [12]. High arousal emotions include higher values for these speech features. for instance the anger and happy emotions are in same arousal state but they are differ in affect(positive and negative valence). This is the key point in valence arousal space i.e how to differentiate the emotions in the same arousal state accurately. Similarly the speech features have lower values in the low arousal space. Hence emotion is a physiological experience of a person's state when interacting with the environment and valence arousal space captures a wide range of significant issues in emotion. As mentioned in the emotional state literature the first five standard archetypal emotions are anger, happy, fear, sad and disgust [11].

1.3.2   Computer Science Based Emotion

One of the most important of outcome of emotion is the speech signal and its features in view of computer. The main source of speech is vibration of vocal cords.The features extracted from the speech signal contains most of the emotion specific information and are mainly classified as two types prosody and spectral. Prosodic features are influenced by vocal fold activity and occurs when we put sounds together in a connected speech [35]. Prosodic features discriminate high arousal emotions with low arousal emotions very accurately [38]. Some examples of prosodic features are pitch,intensity and speech rate etc., Spectral features are influenced by vocal tract activity and are extracted from spectral content of the speech signal [35]. Spectral features discriminate the emotions in the same arousal state by using the information produced in the valence state[60]. Some examples of spectral features are MFCC (Mel Frequency Cepstral Coefficients), LPCC (Linear Prediction Cepstral Coefficients), LFPC (Log Frequency Power Coefficients), formants etc.,

1.4   Applications of Speech Emotion Recognition

The main application of speech emotion recognition is in human computer interaction in web tutorials, where the response of the user to the system is important. Depending on this response the system has to detect the emotional state of the user, basing on this the response of the system changes according to the user emotional state [63]. This can be used as a diagnostic tool for therapist [19]. Speech recognition systems with stressed speech achieve better performance than those trained by normal speech in aircraft cockpits [21]. Speech emotion recognition can also be used in mobile communication and call center applications [39]. It is also useful in car environment systems to detect emotional state of the driver and is given as input to the system to alert him from an accident [63, 17].

1.5   Book Organization

The chapter wise organization of the book is given below.

Chapter 1: Introduction   introduces the representation of the signal and its acoustic features from speech. Different types of emotion in terms of physiology and computer science based are discussed and different types of application in the area of speech emotion recognition are discussed.

Chapter 2: Emotion Recognition Using Prosodic Features   discusses the preprocessing of speech signals like filtering, framing and windowing, before extracting the speech features. Different types of prosodic features like Energy, Zero crossing rate and pitch are reviewed and finally the importance of prosodic features in extracting the emotional specific information is given.

Chapter 3: Emotion Recognition Using Spectral Features   presents procedure for implementing different types of spectral features like Mel Frequency Cepstral Coefficients, Linear Prediction Cepstral Coefficients and formant features and the importance of spectral features in extracting emotions is given.

Chapter 4: Feature Fusion Techniques   discusses different types of feature fusion techniques like Adaptive and non adaptive feature fusion, feature level feature fusion and decision level feature fusion.

Chapter 5: Emotional Speech Corpora   reviews different types of emotional speech databases existed and gives more importance to the databases like Berlin, Spanish databases which are used in this book for practical purpose.

Chapter 6: Classication Models   discusses general classification of classification models like statistical approach, syntactic approach, template matching and neural networks. Different types of classification models which are implemented in this book are discussed elaborately like Linear discriminant analysis, Regularized discriminant analysis, Support vector machine and k-nearest neighbor. Different types of distance measures are discussed briefly.

Chapter 7: Comparative Analysis of Classiers in Emotion Recognition   discusses the experimental results in detail with every aspect which are implemented using berlin and spanish databases over all the classification models discussed in previous chapters. Analysis of results with receiver operating characteristic curves are also discussed.

Chapter 8: Summary and Conclusions   contains the summary of the present work and some important conclusions drawn. This chapter also discusses the scope and future work in the area of speech emotion recognition to improve the performance of the classifiers.

Chapter 2
Emotion Recognition Using Prosodic Features

2.1 Introduction

In computer vision, a feature is a set of measurements. Each measurement contains a piece of information and specifies the property or characteristics of the object. In speech recognition techniques, how the speech signals are produced and perceived by the human is starting point of the research. Human speech communication produces ideas (word sequence) which are made within the speaker brain. These word sequence are delivered by his/her text generator. The general human vocal system is modeled by the speech generator. The speech generator converts the word sequence into speech signal and is transferred to listener through air. At the listener side, the human auditory system receives these acoustic signal and listeners brain starts the processing of signal to understand its content. The speech recognizer modeled by the speech decoder, it decodes the acoustic signal into word sequence. So speech production and speech perception are in inverse processes in the speech recognition application.

When we analyze speech signals, most of them are more or less stable within a short period of time. When we do frame blocking, there may be some overlaps between neighboring frames as shown in Fig. 2.1. Each frame is the basic unit for analysis. The basic approach to the extraction of acoustic features from the speech signal can be summarized as follows:

- Converting the stream of speech signals into set of frames by performing frame blocking. The size of each frame must not be too big and too small. If frame size is too big, we can't extract the time-varying characteristics of the audio signals. And if the frame size is too small, we can't extract the valid acoustic features, then extraction cannot be done. In general, a frame contains several fundamental periods of the given speech signal. Usually the frame size (in terms of sample points) is equal to the powers of 2 (such as 256, 512, 1024 etc.) such that it is suitable for fast Fourier transform.
- For frame blocking we should also consider the duration between the frames. If we want to reduce the difference between neighboring frames, we can allow overlap between them. Usually the overlap is 1/2 to 2/3 of the original frame.

© The Author(s) - SpringerBriefs 2015

K. R. Anne et al., *Acoustic Modeling for Emotion Recognition,*
SpringerBriefs in Electrical and Computer Engineering, DOI 10.1007/978-3-319-15530-2_2

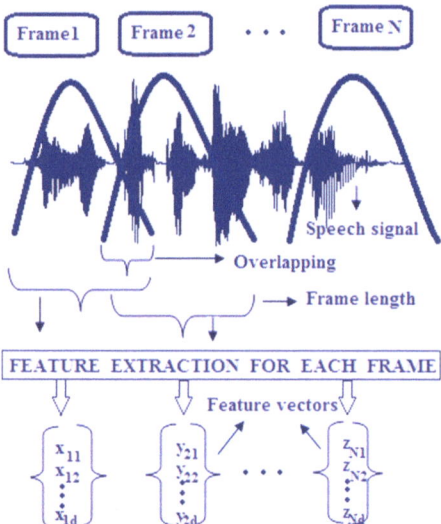

Fig. 2.1 The process of splitting the speech signal into several frames, appling an hamming window for each frame and its corresponding feature vector

The more overlap, the more computation is needed. So we must take care of the distance between two frames.

- After frame blocking, we extract the acoustic features from the audio signals such as zero crossing rate, short time energy, pitch, MFCC etc, can be done by assuming the audio signals within the frame as stationary [87].
- By observing the zero crossing rate and short time energy of the each frame, we have to analyze that particular frame of audio signal is voice speech signal or unvoiced speech signal and keep the voice speech (non-silence frames) for further analysis.

In case of acoustic information, features are mainly classified as temporal and spectral features. Zero Crossing Rate (ZCR), Short Time Energy (STE) are the examples of temporal features. Mel frequency cepstral coefficient (MFCC) and Linear Prediction Cepstral Coefficeints (LPCC) are examples of spectral features.

2.2 Pre-Processing

Speech signals are normally preprocessed before features are extracted to enhance the accuracy and efficiency of the feature extraction process. The modules filtering, framing and windowing are considered as steps under preprocessing.

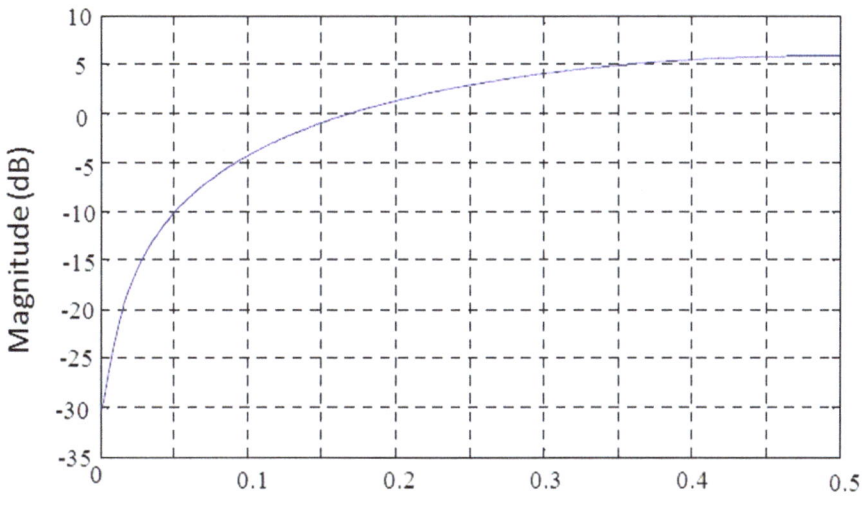

Fig. 2.2 Pre-emphasis filter

2.2.1   *Filtering*

The filtering technique is applied to reduce the noise, which is occured due to the environmental conditions or any other disturbances while recording the speech sample. To reduce the noise effect, filter operations are performed which optimizes the class separability of features [1]. This is done by using the high pass filter.

The main goal of Pre-emphasis is to boost he amount of energy in the higher frequencies with respect to lower frequencies. Mainly boosting is used to get more information from the higher frequencies available to the acoustic model and to improve the recognition performance [2, 7]. This pre-emphasis is done by using a first-order high pass filter as shown in Fig. 2.2.

2.2.2   *Framing*

Frame blocking is converting the stream of audio signal into set of frames and analyzed independently. The original vector of sampled values will be framed into overlapping blocks. Each block will contain 256 samples with adjacent frames being separated by 128 samples. This will yield a minimum of 50 % overlap to ensure that all sampled values are accounted for within at least two blocks. Two hundred and

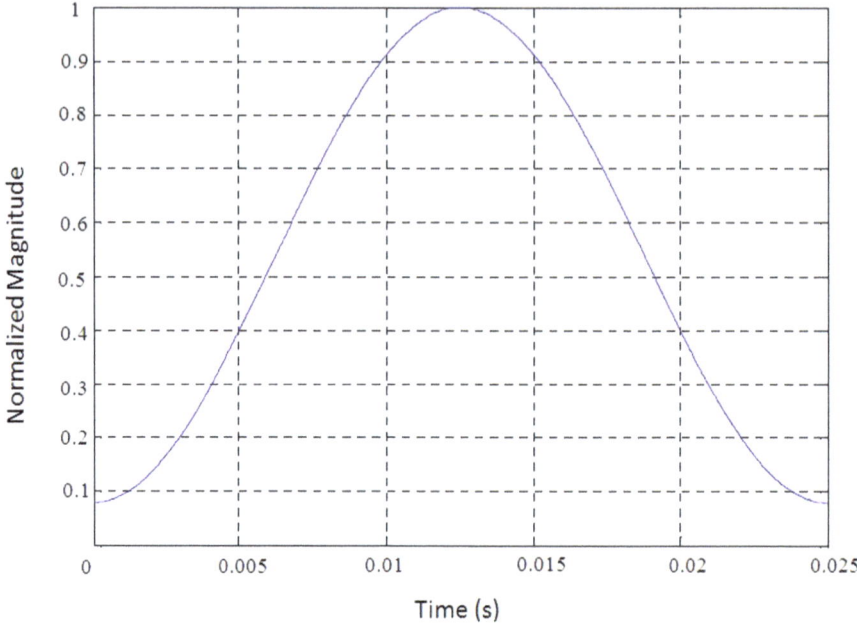

Fig. 2.3 25 ms hamming window fs = 16 khz [1]

fifty six was chosen so that each block is 16 ms. In this step the continuous speech signal is made into frames of N samples, with adjacent frames being separated by M (M < N). Typical values used are M = 100 and N = 256 [1]. Frame blocking is shown in Fig. 2.1.

2.2.3   Windowing

In the window operation, the large input data is divided into small data sets and stored in sequence of frames. While dividing the signal into frames, some of the input data signal may be discontinuous at the edges of the each frame. So a tapered window is applied to each one. The hamming window is used to reduce the spectral leakage in the input data signal.

The generally used window is rectangular window, it is the simplest window. But this window can cause some problems, however, because it abruptly cuts off the signal at its boundaries. These discontinuities create problems when we do Fourier analysis. Therefore it is necessary to keep the continuity of the first and the last points in the frame. For this reason, the Hamming window is used in feature extraction. The 25 ms hamming window is shown in Fig. 2.3.

2.3   Extraction of Prosodic Features

An important module in the design of speech emotion recognition system is the selection of features which best classify the emotions. These features distinguishes the emotions of different classes of speech samples. These are estimated over simple six statistics as shown in Table 2.1.

The prosodic features extracted from the speech signal are Zero Crossing Rate, Short Time Energy and Pitch. Their first and second order differentiation provides new useful information hence the information provided by their derivatives also considered [37]. These features were estimated for each frame together with their first and second derivatives, providing six features per frame and applying statistics as shown in Table 2.1 giving a total of 36 prosodic features.

2.3.1   Zero Crossing Rate

In the context of discrete-time signals, a zero crossing is said to occur if the successive samples have different algebraic signs. The rate at which zero crossings occur is a simple measure of a signal. Zero-crossing rate is measure of number of times the amplitude of the speech signals passes through a value of zero in a given time interval/frame. The calculation of this is as shown in Eq. 2.1

$$ZCR = \frac{1}{N} \sum_{i=0}^{N-1} |sgn(x(i)) - sgn(x_{-1}(i-1))| \qquad \cdot \qquad (2.1)$$

where the value of sgn are

$$
\begin{array}{cc}
1 & x(i) > 0 \\
0 & x(i) = 0 \\
-1 & x(i) < 0
\end{array}
$$

where $x_{-1}(N)$ is a temporary array created to store the previous frame values Eq. 2.1 shows the mathematical formula to calculate feature values using zero crossing rate.

Table 2.1 The statistics and their corresponding symbols [35]

Statistics	Symbol
Mean	E
Variance	V
Mininum	Min
Range	R
skewness	Sk
kurtosis	K

The function of sgn in the equation is to assign the normalized value $[-1, 0, 1]$ based on the range of input variable value. Since high frequencies imply high zero crossing rates, and low frequencies imply low zero-crossing rates, there is a strong correlation between zero crossing rate and energy distribution with frequency [74]. A reasonable generalization is that if the zero-crossing rate is high, the speech signal is unvoiced, while if the zero-crossing rate is low, the speech signal is voiced.

2.3.2 Short Time Energy

The amplitude of the speech signal varies with time. Generally, the amplitude of unvoiced speech segments is much lower than the amplitude of voiced segments. The energy of the speech signal provides a representation that reflects these amplitude variations . Short-time energy can define in Eq. 2.2

$$STE = \frac{1}{N} \sum_{i=0}^{N-1} |X(n)|^2 \tag{2.2}$$

where N describes the total number of samples in a frame or a window. $X(n)$ is a speech signal in a frame. A reasonable generalization is that if the Short time energy is high, the speech signal is voiced, while if the Short time energy is low, the speech signal is unvoiced. Based on zero crossing rate and short time energy, voiced sounds are identified. We can extract the following features from the identified voice speech signal.

2.3.3 Pitch

Pitch is the fundamental frequency of audio signals, which is equal to the reciprocal of the fundamental period [75]. This is mainly explained in terms of highness or lowness of a sound. Pitch in reality can be defined as the repeat rate of a complex signal, i.e., the rate at which peaks in the autocorrelation function occur. The three main difficulties in pitch extraction arise due to the following factors:

- Vocal cord vibration does not necessarily have complete periodicity, especially at the beginning and end of the voiced sounds.
- From speech wave, vocal cord source signal can be extracted but its extraction is difficult if it has to be extracted separately from the vocal tract effects.
- The fundamental frequency possesses very large dynamic range.

Above three viewpoints are very much used for the research on pitch extraction recently. The main point is being able to extract quasi-periodic signal's periodicity in a reliable manner. Another is how to correct the pitch extraction error owing to the disturbance of periodicity. The other is how to remove the vocal tract (formant)

Table 2.2 Classification of major Pitch extraction methods and their major principle features [20]

Classification	Pitch extraction method	Principal features
1. Waveform processing	Parallel processing method	Uses majority rule for pitch periods extracted by many kinds of simple waveform peak detectors
	Data reduction method	Removes superfluous waveform data based on various logical processing and leaves only pitch pulses
	Zero crossing count method	Utilizes iterative pattern in waveform zero crossing rate
2. Correlation processing	Autocorrelation method	Employs autocorrelation of waveform. Applies center and peak clipping for spectrum flattering and computation simplification
	Modified correlation method	Utilizes autocorrelation function for residualsignal of LPC analysis. Computation is simplified by LPF and polarization
	SIFT (simplified Inverse filter tracking) algorithm	Applied LPC analysis for spectrum flattering after down-sampling of speech wave. Time resolution is recovered by interpolation
	AMDF method	Uses average magnitude differential function (AMDF) for speech or residual signal for periodicity detection
3. Spectrum processing	Cepstrum method	Separates spectral envelope and fine structure by inverse Fourier transform of log power spectrum
	Period histogram method	Utilizes histogram for harmonic components in spectral domain. Pitch is decided as the common divisor for harmonic components

effects. Double-pitch and half-pitch are the two main classifications of major errors in pitch extraction. So as the name suggests the double-pitch errors are those which occur while extracting the frequencies which are twice as large as actual value and similarly the half-pitch errors occur due to extraction of half-value of the original fundamental frequency. The employed extraction method plays a major role in deciding the tendency toward which the error is most apt to occur. Precise information regarding all the major pitch extraction methods are given in the Table 2.2. Most of the works uses Auto correlation method to detect pitch.

Pitch Detection using Autocorrelation Method:

Autocorrelation function is used to estimate pitch, directly from the waveform. xcorr function is used to estimate the statistical cross-correlation sequence of random process and is given by

$$R[m] = E(x[n+m]x[n]) = E(x[n]x[n-m]) \tag{2.3}$$

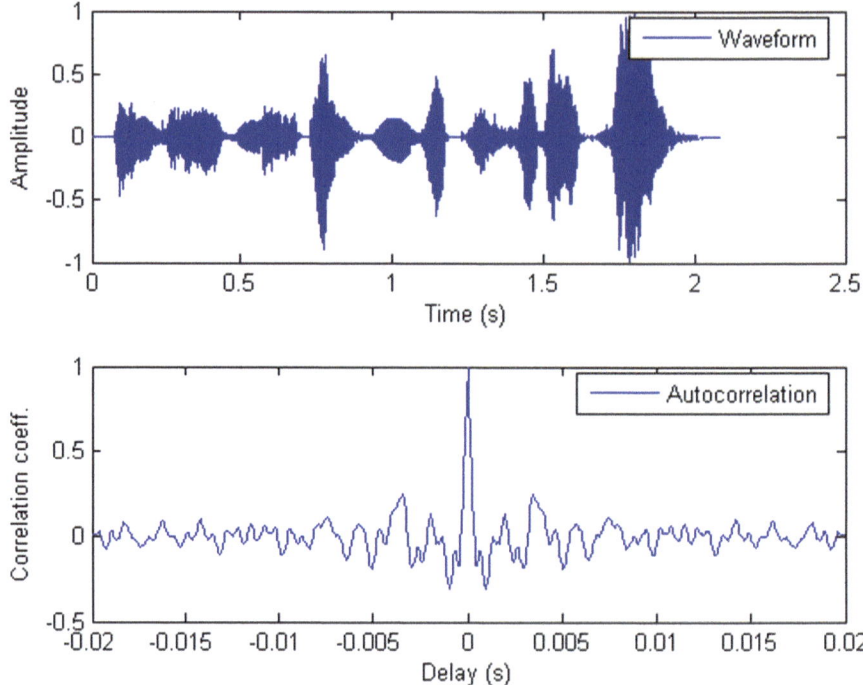

Fig. 2.4 Waveform and autocorrelation function in pitch estimation

When x and y are not the same length then the shorter vector is zero padded to the length of the longer vector. If x and y are length N vectors (N > 1), c = xcorr(x,y) returns the cross-correlation sequence in a length 2*N−1 vector [45, 75]. The autocorrelation function is given by

$$R(m) = \frac{1}{N} \sum_{n=0}^{N-m-1} (x[n+m]x * [n])m \geq 0 \tag{2.4}$$

Where x(n) is a speech signal, n is time for discrete signal, m is a lag number. If x(n) is similar with x(n + m), then R[m] has a large value. Whenever x(n) has a period of P, then the value of R[m] has peaks at m = lP where l is an integer [25, 31]. Here we need to estimate R[m] from N samples. The empirical autocorrelation function is given by $R(m) = \frac{1}{N} \sum_{n=0}^{N-m-1} (w[n+m]x[n+m]w[n]x * [n])$, where w[n] is a window function of length N. The peaks of autocorrelation function and waveform of a speech signal as shown in Fig. 2.4.

We can estimate the fundamental frequency by using autocorrelation function, peaks at delay intervals corresponding to the normal pitch range in speech, say 2 ms (= 500 Hz) and 20 ms (= 50 Hz). Figure 2.4 shows the pitch tracking using autocorrelation method. Pitch is very low for male voice and very high for female voice [69].

2.4   Importance of Prosodic Features

The features extracted from the speech signal plays a major role in identifying emotion and are capable of detecting an exact emotion of an unknown speaker who is not visible during training period. But the detection of feature set which best classify the emotion is a complex task [60].

Some physiological and psychological changes occurs due to each and every emotion. For instance happiness occurs when we stood stood first in competitive examination. This causes some changes in characteristic of speech also like amplitude, frequency and speech rate etc. These are nothing but prosodic features [38]. Because of this reason, emotion identification systems uses prosodic features for many years. Many works in the literature deals with these prosodic features in identifying the emotion [9, 24, 62]. A detailed summary of these works is given in [18]. Most of the emotional information is obtained by using these prosodic features and are estimated over short term speech segments by using simple statistics like minimum, maximum, mean, variance, skewness and kurtosis etc., As shown in Fig. 1.2 prosodic features discriminate high arousal emotions to low arousal emotions more accurately.

Chapter 3
Emotion Recognition Using Spectral Features

3.1 Introduction

This chapter introduces the concept of Spectral features. The procedure for implementing different types of spectral features is given clearly. The spectral features outperform the prosodic features and the importance of spectral features in extracting emotion is also given in the last section of this chapter.

3.2 Extraction of Spectral Features

Spectral features are most efficient in order to extract correct emotional state of the speech sample [38]. Mel Frequency Cepsstral Coefficients(MFCC), Linear Prediction Cepstral Coefficients (LPCC), formants are some examples of Spectral features. Similar to prosody statistics, spectral statistics are calculated using the statistics as shown in the Table 2.1. The most frequently extracted MFCC features are 18 and the number of filter banks used are 24. The extracted 18 MFCC Coefficients and their first and second derivatives are estimated for each frame giving a total of 54 spectral features. The statistics as mentioned earlier are applied to these 54 values so totally $54 \times 6 = 324$ different features are calculated.

3.2.1 MelFrequency Cepstral Coefficients (MFCC)

MFCC are the most widely used spectral representation of speech. MFCC features represents the short term power spectrum of a speech signal, based on a linear cosine transform of a log power spectrum on a non linear melscale of frequency. The procedure for implementing MFCC is shown in [57, 70]. MFCC is based on human hearing perceptions which cannot perceive frequencies over 1000 Hz or 1 KHz. In other words, MFCC is based on known variation of the human ear's critical bandwidth with frequency. MFCC has two types of filter which are spaced linearly

© The Author(s) - SpringerBriefs 2015
K. R. Anne et al., *Acoustic Modeling for Emotion Recognition*,
SpringerBriefs in Electrical and Computer Engineering, DOI 10.1007/978-3-319-15530-2_3

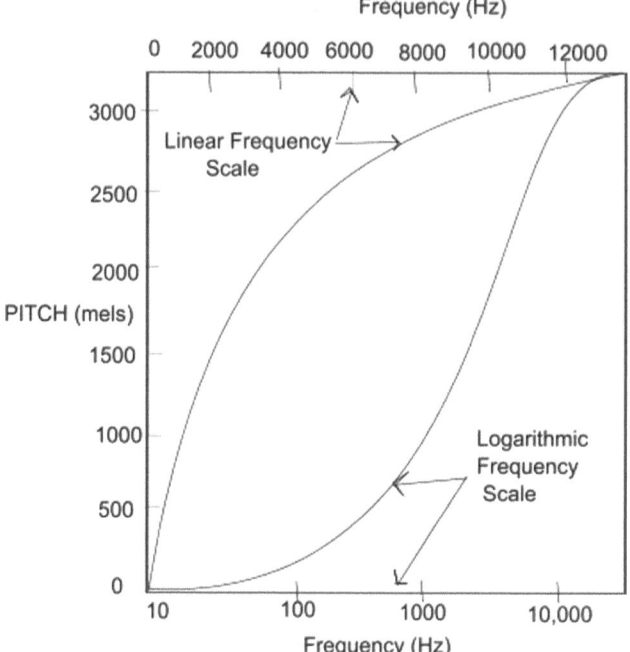

Fig. 3.1 Relation between the perceived pitch and frequency

at low frequency below 1 KHz and logarithmic spacing above 1 KHz [29, 66, 34]. A subjective pitch is present on Mel Frequency Scale to capture important characteristic of phonetic in speech. It turns out that humans perceive sound in a highly nonlinear way. Basic parameters like pitch and loudness highly depend on the frequency, adding weight to components at lower frequencies.

In Fig. 3.1 the behavior relating the perceived pitch to the physical frequency is illustrated. The pitch associated with a tone is measured on the so-called mel-scale (By definition 1000 mels correspond to the perception of a sinusoidal tone at 1 KHz, 40 dB above the hearing threshold).

The Fig. 3.2 clearly shows that the perceived pitch increases all the more slower as we go to higher frequencies. Essentially we observe a logarithmic increase that is illustrated by the almost linear curve (with respect to a logarithmic scale) in Fig. 3.1 at high frequencies. MFCCs extensively use this property and add weight to lower frequencies, because more discriminative information can be found there. After frame blocking and windowing the MFCC consists of several computational steps. Each step has its function and mathematical approaches as discussed briefly in the following:

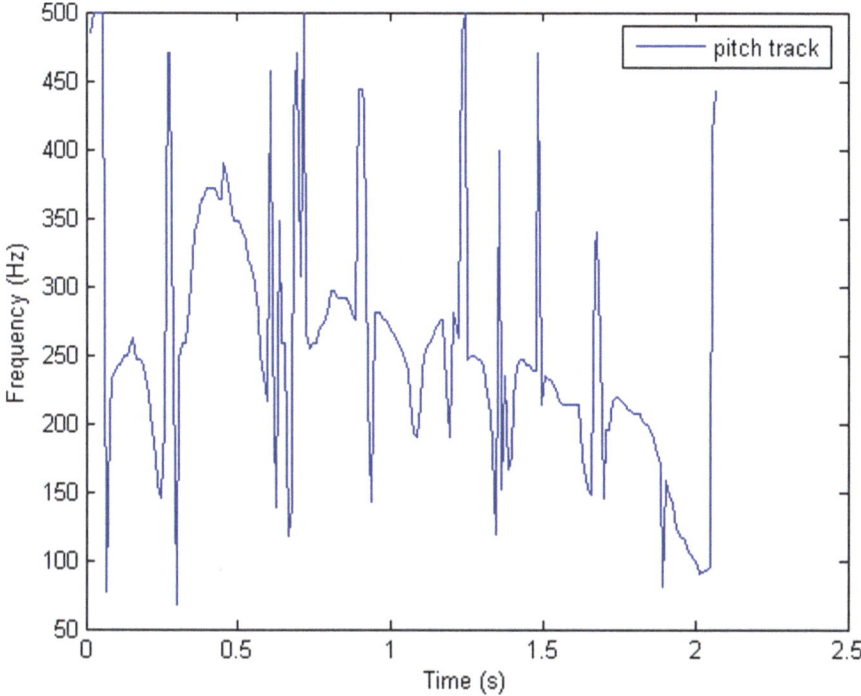

Fig. 3.2 Pitch tracking using autocorrelation method. The estimated pitch is 286 Hz

Step 1: Fast Fourier Transform (FFT)

In order to analyze the audio data in frequency domain, Fourier transform is applied for the input data. Fourier transform can be applied using various methods like Discrete Fourier Transform (DFT), Fast Fourier Transform (FFT), etc [68].

As from the available methods, FFT is capable of generating results quickly. When the input data is divided into frames, the values of each frame is converted into frequency domain using FFT. For the frequency domain algorithms, as the window size changes, the execution time and memory requirements also change. The mathematical equation to calculate the FFT values explained as follows [65, 43]

$$X_k = \sum_{n=0}^{N-1} x_n e^{-j2\pi k \frac{n}{N}} \; where \; k = 0, 1, 2, ...N - 1 \tag{3.1}$$

Consideration of only absolute values(frequency magnitudes) of X_k which are complex numbers integral. The interpretation of the resulting sequence X_k is as follows.

Positive frequencies $0 \leq f < F_s/2$ correspond to values $0 \leq n < N/2 - 1$, negative frequencies $Fs/2 < f < 0$ correspond to $N/2 + 1 \leq n \leq N - 1$ here, F_s denotes the sampling frequency.

Step 2: Mel-Scaled Filter Bank and Log Processing
The frequencies range in FFT spectrum is very wide and voice signal does not follow the linear scale. Magnitude of the filter frequency response is used to get the log energy of that filter. Here a set of 20 triangular bandpass filters are used. Each filter's magnitude frequency response is triangular in shape and equal to unity at the center frequency and decrease linearly to zero at center frequency of two adjacent filters [43]. The sum of the filtered spectral components is the output of each filter. After that the Eq. 3.2 is used to compute the Mel for given frequency f in Hz:

$$m = 2595 log_{10}\left(\frac{f}{700} + 1\right) = 1127 log_e\left(\frac{f}{700} + 1\right) \tag{3.2}$$

and the inverse:

$$f = 700(10^{m/2595-1}) = 700(e^{m/1127-1}) \tag{3.3}$$

An alternate formula, not depending on choice of log base is

$$m = (1000/log(2))(log(f/1000 + 1)) \tag{3.4}$$

Similar to effects in the humans subjective aural perception, the Mel-frequency is proportional to the logarithm of the linear frequency. The Mel-frequency scale is linear frequency spacing below 1000 Hz and a logarithmic spacing above 1000 Hz.

According to Mel scale, a set of triangular filters are shown in the Fig. 3.3 that are used to compute a weighted sum of filter spectral components, so that the output of process approximates to a Mel scale. The reasons for using triangular band pass filters are:

- Smooth magnitude spectrum such that the harmonics are flattened in order to obtain the envelop of the spectrum with harmonics. This indicates that the pitch of a speech signal is generally not presented in MFCC. As a result, a speech recognition system will behave more or less the same when the input utterances are of the same timbre but with different tones/pitch.
- Sizes of the features involved are reduced.

Step 3: Discrete Cosine Transform
A Mel frequency cepstrum coefficient (MFCC) is obtained by the conversion of the log Mel spectrum back to time domain. The set of coefficient is called acoustic vectors. Therefore, transformation of each input utterance into a sequence of acoustic vector. For the given frame analysis, the local spectral properties of the signal can be represented in a good way by using the cepstral representation of the speech spectrum. Conversion of the Mel spectrum coefficients (and so their logarithm) to the time domain using the DCT is possible as they are real numbers [43, 7]. Here, DCT on the 20 log energy E_k is applied and obtained from the triangular band-pass filters to have L Mel-scale cepstral coefficients [43]. The formula for DCT is given by the Eq. 3.5

$$C_m = \sum_{k=1}^{N} E_k.cos\left[m * \left(k - \frac{1}{2}\right)\frac{\pi}{N}\right] \quad m = 1, 2, ..., L \tag{3.5}$$

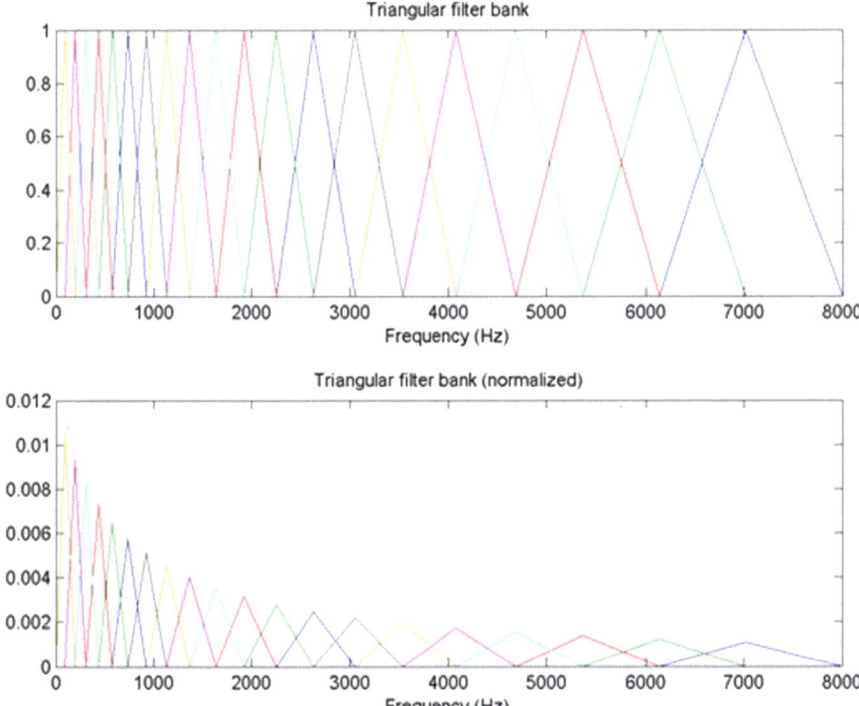

Fig. 3.3 An example of mel-spaced filterbank [43]

Here we set $N = 20$ and $L = 12$, where the number of log spectral coefficients N and the number of Melscale cepstral coefficients L. The obtained features are referred to as the Mel-scale cepstral coefficients or MFCC because these are similar to cepstrum. The main feature for speech recognition can be MFCC alone. By adding of the log energy and by performing delta operation, performance can be increased.

Step 4: Delta Cepstrum and Delta Energy
The velocity and acceleration of (energy+MFCC) are shown by the time derivatives of (energy+MFCC), this is an advantageous new feature. Over the time, features related to the change in cepstral features are to be added. 12 cepstral features, 13 delta or velocity features and 13 double delta or acceleration features i.e. total of 39 features are used [43]. The Eq. 3.6 represents the energy in a frame for a signal x (from time sample t1 to time sample t2).

$$Energy = \sum X^2[t] \tag{3.6}$$

Corresponding cepstral or energy feature, change between frames in the Eq. 3.6 is represented by each of the 13 delta features, while the change between frames in the corresponding delta features is given by Eq. 3.7 which represents each of the double

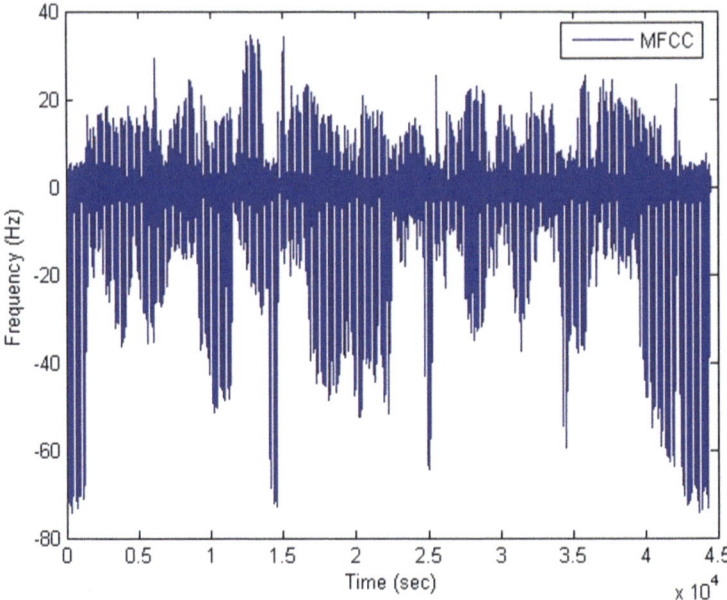

Fig. 3.4 Mel frequency cepstral coefficients

delta features

$$d(t) = \frac{c(t+1) - c(t-1)}{2} \tag{3.7}$$

After adding energy, delta and double delta features to the 13 cepstral features, totally 39 MFCC features are extracted. And one of the most useful fact about MFCC features is that the cepstral coefficients tend to be uncorrelated. This turns our acoustic model much simpler.

MFCC is the most widely used spectral representation of speech. MFCC parameters are calculated by taking the absolute value of the FFT, warping it to a Mel frequency scale, taking the DCT of the log-Mel spectrum and returning the first 13 coefficients. The Fig. 3.4 is a plot of the MFCC. The function requires the following parameters: signal, sampling frequency, window type, number of coefficients, number of filters in the filter bank, length of a frame and the frame increment.

3.2.2   Linear Prediciton Cepstral Coefficients (LPCC)

Linear Prediction is a well known method for removing redundancy in a speech signal [5]. The prediction is done in two separate stages [4]. one prediction is based on short time spectral envelope of speech and another prediction is based on periodic nature of spectral fine structure [3]. Most of the emotion specific information can be obtained by using these Linear prediction coefficients. spectral envelope of speech is determined by the frequency response of the vocal tract and the spectrum of glottal pulse. The spectral fine structure is determined mainly by the pitch period [5]. The structure of unvoiced speech is random so it can not be used for prediction.

The predictor can be characterized in the z-transform notation as [5]

$$P_s(z) = \sum_{k=1}^{p} a_k z^{-k} \tag{3.8}$$

where Z^{-1} represents a delay of one sample interval and $a_1, a_2, ...a_p$ are p predictor coefficients. These are used to predict the speech sample.

The Linear predictive coding gets its name from the fact that it predicts the current sample as a linear combination of its past p samples.

$$\tilde{x}[n] = \sum_{k=1}^{p} a_k x[n-k] \tag{3.9}$$

the difference between actual and predicted speech sample is known as an error and is given in Eq. 3.10

$$e(n) = x[n] - \tilde{x}[n] = x[n] - \sum_{k=1}^{p} a_k x[n-k] \tag{3.10}$$

where $x[n]$ is the original speech signal and $\tilde{x}[n]$ is the predicted speech signal. To estimate the predictor coefficients from a set of speech samples, we use short term analysis technique. Let $x_m[n]$ is speech segment of sample m.

$$x_m[n] = x[m+n] \tag{3.11}$$

The short time prediction error for that segment can be defined as the sum of squared differences between the actual and predicted speech samples and the error should be minimized to get its LPC Coefficients.

$$E_m = \sum_{n} e_m^2[n] = \sum_{n} \left(x_m[n] - \sum_{j=1}^{p} a_j x_m[n-j] \right)^2 \tag{3.12}$$

where n is the number of samples in an analysis frame. To get the LP coefficients, take the derivative of the Eq.˜refs5 with respect to a_i and equate it to 0. we obtain

$$\frac{\partial E_m}{\partial a_k} = 0 \, for \, k = 1, 2, ..., p$$

the cepstral coefficients can be obtained from the LPC coefficients by the following recursion [25]

$$h(n) = \begin{cases} 0 & n < 0 \\ \ln G & n = 0 \\ a_n + \sum_{k=1}^{n-p} \left(\frac{k}{n}\right) h[k] a_{n-k} & 0 < n \leq p \\ \sum_{k=n-p}^{n-1} \left(\frac{k}{n}\right) h[k] a_{n-k} & n > p \end{cases} \tag{3.13}$$

While there are a finite number of LPC Coefficients the number of cepstrum coefficients is infinite. speech recognition researchers have shown that a finite number is sufficient. 12–20 depending on the sampling rate and whther or not frequency warping is done or not [25].

3.2.3 Formant Features

As the sound is propagated through vocal tract and nasal tract tubes, the frequency spectrum is shaped by the frequency selectivity of the tube. This effect is similar to the resonance effect [53]. These resonant frequencies of the vocal tract tube are called formant frequencies or formants and depends upon shape and dimension of the vocal tract and they convey differences between different sounds. As the shape of the vocal tract varies, the spectral properties of the speech signal varies with time and are represented graphically by using a spectrogram. The horizontal dimension corresponds to time and vertical dimension corresponds to frequency. The dark regions of the spectrogram corresponds to energy. Thus the formant frequencies corresponds to the dark regions of the spectrogram. These are very useful features but are not widely used because of the complexity in estimating them [25].

One way of getting formant frequencies is by computing the roots of a $p^t h$ order polyniomaial [3]. The complex roots of a polynomial with real coefficients are computed by using some standard algorithms [52]. Each complex root can be represented as [25]

$$z_i = \exp\left(-\pi b_i + j2\pi f_i\right) \tag{3.14}$$

where b_i and f_i are the formant frequency and band width of the ith root. Complex roots are sorted by increasing f and discarding real roots. The formant candidates are (f_i, b_i) The formant trackers discard the roots whose bandwidths are greater than a thereshold say 200 Hz [78]

Another method is to find the peaks on a smoothed spectrum which is obtained through LPC Analysis [40]. The advantage of this method is, we can always compute the peaks and is more efficient than extracting complex roots of a polynomial [58]. The first three formants are used for formant synthesis since they allow sound classification where as the higher formants are speaker dependent [25].

3.3 Importance of Spectral Features

Some confusion is generated in recognizing emotions through prosodic features. Because they differentiate the emotions in the different arousal states more accurately than that of emotions in same arousal states. For instance as shown in Fig. 1.2 the confusion level among the emotions in the same arousal state like excited and angry

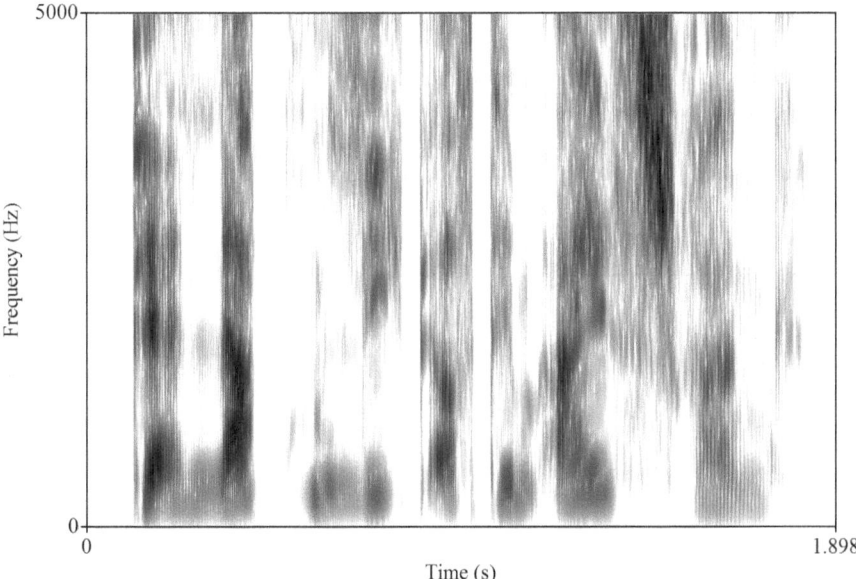

Fig. 3.5 Spectrogram of one sample speech signal

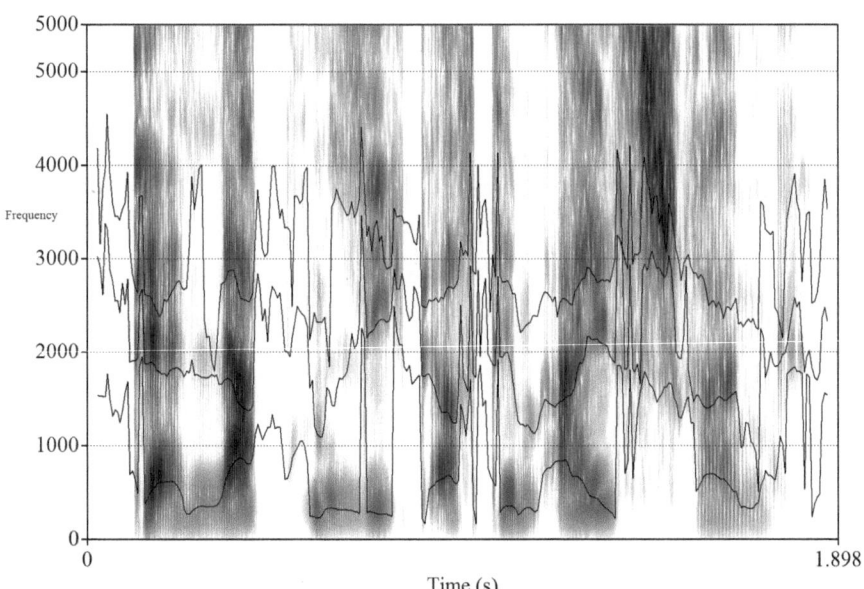

Fig. 3.6 Spectrogram and its corresponding smoothed formants

is very large but they are able to discriminate anger from sad which are on different emotional states.

Because the spectral features are influenced by vocal tract and prosodic features are influenced by vocal fold they both provide different types of emotional specific information [38]. Many works in the literature deals with the relation between spectral features and the emotion [44, 61] and they give good results in identifying the emotions than that of with prosodic features. The combination of these two features is called feature fusion which is also give better results. The results with individual features and fused features are given in further chapters.

Chapter 4
Feature Fusion Techniques

4.1 Introduction

Researcher introduced biometric system to improve the security because nowadays, the world is becoming more insecure. The term Biometrics is derived from the Greek composite words bio and metric. Bio means life and metric means to measure. Biometrics basically is a measure and analysis of human physiological and behavioral characteristics to identify and verify the person. First biometric are introduced in China in fourteenth century [56]. In 1890, Alphonse Bertillon, an anthropologist introduced biometric in Europe to identify criminals. In the past two decades, there has been an explosion in the biometric industry, where a variety of biometric techniques have been introduced in the market [46]. Some of these biometric techniques measure the human physiological characteristics (such as fingerprints, speech, eye retinas and irises, facial patterns, DNA and hand measurements [76] etc) and behavioral characteristics (such as speaker, signature and keystroke etc).

The biometric systems are mainly classified into

- Unibiometric system: This biometric system uses a single instance or a single sample, a single representation and a single matcher for a verification, identification and recognition.
- Multi biometric system: This biometric system uses a more than one biometric identifier with multiple instances or multiple samples (multiple thumb impressions of a finger, multiple images of a face in a video, multiple voice samples), multiple representations of a single input, multiple matchers of a single representation, or any one combination (for example, fingerprint and voice of the same person).

One of the important benefits in multi biometric system is if one input is highly noisy, then the other input might be helpful to make an overall reliable decision. Other limitations of Unibiometrics are reliability of sensed data, non/univesality and lack of performance when compared to multi biometrics.

© The Author(s) - SpringerBriefs 2015   27
K. R. Anne et al., *Acoustic Modeling for Emotion Recognition*,
SpringerBriefs in Electrical and Computer Engineering, DOI 10.1007/978-3-319-15530-2_4

4.2   Multi Modal Feature Fusion

To avoid these limitations, the researchers and developers come up with different types of Multi biometric systems. In multi biometric, we have more than one input sample [56]. The multi biometric systems available in the market are:

- Multi Sensor Biometric Systems: As the name suggest, this biometric system consists of multiple sensors to capture the data/sample i.e one sensor capture the data/sample once (here we have multiple sensors so we have multiple data/samples) and then combine them to improve the accuracy and security [46]. For instance, 2D cameras as well as infrared sensor can be used in this biometric system to capture the images of an individual in different poses and different illumination conditions. For example, optical and solid-state sensors can be used to capture the fingerprints of the same person and same finger.
- Multi Algorithm Biometric Systems: This type of biometric system does not require any extra sensors and devices. This system needs one sensor and acquires a single sample [50]. In this system, to improve the security, we use more than one algorithm to process this single sample and fused these result to obtain the overall result. The main disadvantage with this system is complexity is high.
- Multi Sample or Multi instance Biometric Systems: In this type of biometric systems, the multiple samples are collected from the same sensor and the same algorithm processes the each sample, finally fuse these multiple samples to enhance the security [47]. The 2D cameras can be used in this biometric system to capture the images of an individual in different poses. For example, we have 10 fingers thus, this biometric technology can make use of impressions of all the 10 fingers and processes these with the one algorithm and fuse the result to get the overall result.
- Multimodal Biometric Systems: This type of biometric system uses different biometric sources (physically uncorrelated samples). Except multimodal, all the other types of biometric systems use single biometric source [56]. In this biometrics, we can select the modalities which are better for our application.

For instance voice patterns and face recognition used together or iris patterns with fingerprints and so on.

4.2.1   Adaptive and Non Adaptive Feature Fusion

Generally, fusion techniques are distniquished as adaptive and non adaptive. Adaptive (quality-based) fusion attempts to change the decision weight as a function of the signal quality measured. The idea is to give higher weights to the decision with the higher quality. For instance, in biometric person recognition, if the facial image is corrupted by bad lighting condition, the output of the speech system may be weighed more, and when speech system is corrupted by noise, the output of the visual facial

system may be weighted more [50, 56]. The quality of an incoming input is measured
by using the quality measures.

Quality measures assess the quality of incoming signal or sample depends upon
the application. Quality measures examples in case of face recognition application are face detection reliability, presence of glasses, lighting conditions, contrast,
etc.Quality measures examples in case of speech recognition application are signal
to noise ratio, echo and gaps in speech. More than one quality measure should be
considered for higher performance in practical.

The non-adaptive fusion does not consider any quality measures. A non-adaptive
linear fusion classifier will take the form of

$$y_{com} = \sum_{i=1}^{N} w_i y_i + w_0 \tag{4.1}$$

where $w_i R$ is the weight associated to the output y_i and w_0 is a bias term. In contrast,
the adaptive fusion classifier would be computed as shown in Eq. 4.2

$$y_{com} = \sum_{i=1}^{N} w_i(q) y_i + w_0(q) \tag{4.2}$$

where $w_i(q)$ changes with the quality signal q. For the sake of simplicity, assume
for now that $q = q_1, ..., q_N$, where q_i is the quality measure of the ith modality.
In general $w_i(q)$ could be of any functional form. However, we shall assume that
weights vary linearly as a function of quality, that is shown in Eqs. 4.3 4.4

$$w_i(q) = \sum_i w_i^{(2)} q_i + w_i^{(1)} \tag{4.3}$$

$$w_0(q) = \sum w_0^{(0)} q_i + w_0^{(0)} \tag{4.4}$$

substituting the Eqs. 4.3 and 4.4 in 4.2

$$y_{com} = \sum_{i=1}^{N} (w_i^{(2)} q_i + w_i^{(1)}) y_i + \sum_{i=1}^{N} (w_0^{(0)} q_i + w_0^{(0)}) \tag{4.5}$$

$$y_{com} = \sum_{i=1}^{N} w_i^{(2)} q_i y_i + \sum_{i=1}^{N} w_i^{(1)} y_i + \sum_{i=1}^{N} w_0^{(0)} q_i + w_0^{(0)} \tag{4.6}$$

where we note that $w_i^{(2)}$ is the weight associated with pairwise element $y_i.q_i$,$w^{(1)}$ is
the weight associated to y_i, and $w^{(0)}$ is the weight associated to q_i. In multimodal
fusion, the data collected from different sources at different levels are integrated.
Fusions are of different levels, they are namely:

- Feature Level Fusion
- Decision Level Fusion

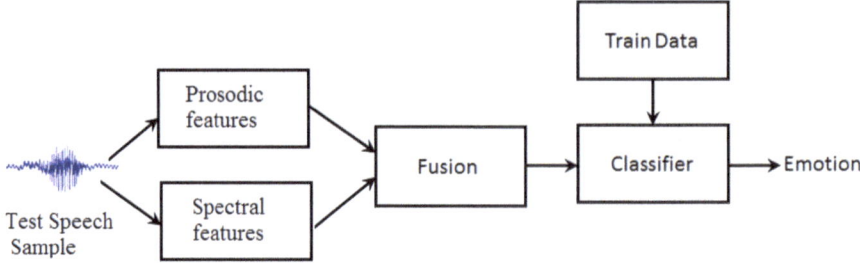

Fig. 4.1 Schematic description of fusion at feature level

4.2.2 Feature Level Feature Fusion

This is also called Feature Fusion in pre-processing. Feature fusion means fusion of features after feature extraction. In this level, the feature vectors extracted from the input speech sample are integrated. If the two different feature vectors of the same speech sample (one is from prosodic features, the other is from spectral features) are extracted, then it is possible to combine and create a new and more reliable feature vector from these two vectors. However, if the samples are of different types (face and voice data), the feature vectors can be concatenated into a new and more detailed feature vector. Figure 4.1 shows the fusion of speech features at feature level.

Fusion at feature level is difficult in practice because concatenating two feature vectors may result in a feature vector with very large dimensionality leading to the 'curse of dimensionality' problem [55].

4.2.3 Decision Level Feature Fusion

This is also called Feature Fusion in Model building. In decision level multimodal fusion, individual subsystems have its own decision. In the fusion process, multiple decisions are fused into single decision. Figure 4.2 shows the fusion at decision level. For decision level fusion, we can use either multiple samples for the same type of sensors or multiple sample from different types of sensors [56]. Here multiple samples from the same type of sensor information namely acoustic features are processed independently and finally the decisions of the appropriate classifiers are fused as shown in Fig. 4.2

c1	c2	f1	f2	f3	f4	f5	f6	f7	f8
0	0	0	0	0	0	0	0	0	0
0	1	0	0	0	0	1	1	1	1
1	0	0	0	1	1	0	0	1	1
1	1	0	1	0	1	0	1	0	1

c1	c2	f9	f10	f11	f12	f13	f14	f15	f16
1	1	1	1	1	1	1	1	1	1
0	1	0	0	0	0	1	1	1	1
1	0	0	0	1	1	0	0	1	1
1	1	0	1	0	1	0	1	0	1

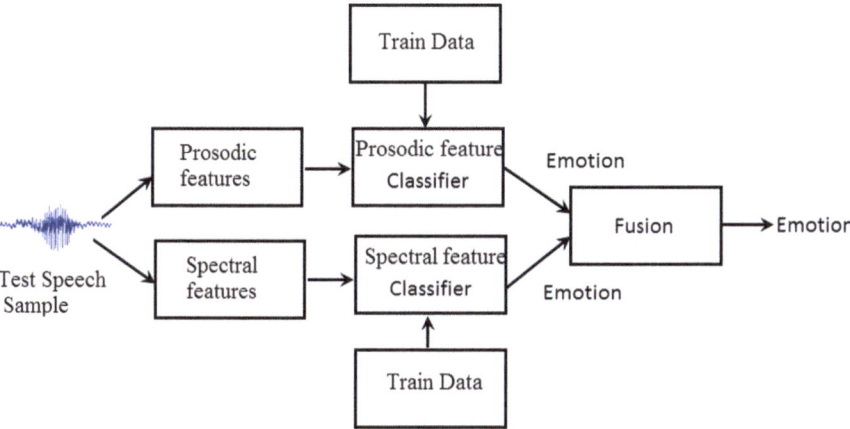

Fig. 4.2 Schematic description of fusion at decision level

Decision Level Fusion of Two Classifiers

For two classifier fusion, we have 16 decision fusion rules as shown in the table above. In two classifiers, we have only acceptance (match) and rejection (not match). Here "1" represents the acceptance and "0" represent the rejection [72].

In these 16 decision rules, the most commonly used rules are f2 ("AND" rule) and f8 ("OR" rule). AND rule accepts,if both classifiers accepts. OR rule accepts,if either one of the classifier accepts. The f9 is "NAND" rule. This rule accepts, if both the classifiers reject. The f1 rule simply rejects irrespective of individual classifiers decisions [72]. Similarly, the f16 rule gives acceptance irrespective of the individual classifiers decisions. In most cases, expect "AND" and "OR" these rule does not logically make sense and leads to poor performance. In multimodal verification system, "AND" and "OR" rule is a simplest method to combine the decision output of different multimodal subsystems. The output of "AND" rule is the match when the both subsystems input samples matches with the train data template. By using "AND", the False Acceptance Rate (FAR) of the multimodal biometric system is lower than the FAR of individual biometric subsystem [55]. The output of "OR" rule is the match of which one of the subsystem input sample matches the train data template. By using "OR", the False Reject Rate (FRR) is greater than the FRR of individual subsystem [55].

But this "OR" rule is not possible in emotion recognition applications. Because if the two visual and acoustic subsystems has different emotions, if we apply "OR" rule, we can't decide which emotion is the final output. And another problem with this "AND" and "OR" rule is, in real world scenario, acoustic information is present in bursts based on the emotion of the driver, where as visual information is present throughout the driving. Because of this limitation, we could not considered "AND" and "OR" rule for driver emotion recognition scenario.

4.2.3.1 Decision Level Fusion of Multiple Classifiers

The most common and simplest rule derived from the sum rule for decision level fusion is majority voting rule. In multimodal verification system, input samples are assigned to the subsystems and identifies the majority of the subsystems agrees the match or not [55]. If the input samples are R, then atleast k are identified as matched then final output of the decision level is "match". k is shown in Eq. 4.7. Atleast k matches should agree that identity.

$$k = \begin{cases} \frac{R}{2} + 1 & if \ R \ is \ even \\ \frac{R+1}{2} & otherwise \end{cases} \tag{4.7}$$

The major drawback in majority voting is all the subsystems are treated or weighted equally.

Chapter 5
Emotional Speech Corpora

5.1 Introduction

Emotion databases are divided into acted/non-acted, induced and naturalistic databases [15]. Simulated databases are also called actor based databases for instance they are collected from experienced professional actors. The actors are given some sentences and they are asked to express these sentences in different emotions. These are called full blown emotions and ar more expressive than real ones [17, 32]. Induced data bases are also called Elicited data bases and are collected by creating an artificial emotional situation. in this case the speakers are made to involve in the conversational situation with the anchor. With out the knowledge of the speaker that they have been recording the data bases are more natural and realistic ones than the simulated ones. Natural emotions are some what difficult to record because collection of wide range of emotions naturally is difficult. Some of the instances of naturalistic emotions are collected from call center conversations, cockpit recordings during unnatural conditions, patient doctor conversations and so on.

5.2 Overview of Emotional Speech Databases

In recent years, major speech processing labs world wide are trying to develop efficient algorithms for emotion speech synthesis and speech emotion recognition [73]. To achieve this the pre requisite is the collection of emotional speech databases. There are 32 emotional speech databases are created. In that some are publicly available and some are licensed versions. These databases are clearly given in [73]. some of them are English, German, Japanese, Dutch, Spanish, Danish emotional speech database and so on. Here we are using two emotional speech databases mainly and are given as follows.

© The Author(s) - SpringerBriefs 2015   33
K. R. Anne et al., *Acoustic Modeling for Emotion Recognition*,
SpringerBriefs in Electrical and Computer Engineering, DOI 10.1007/978-3-319-15530-2_5

5.3   Berlin Emotional Speech Database

Berlin emotional speech database is developed by the Technical University, Institute for Speech and Communication, Department of Communication Science, Berlin [10]. It has become one of the most popular databases used by researchers on speech emotion recognition, thus facilitating performance comparisons with other studies. Five actors and five actresses have contributed speech samples for this database, it mainly has ten German utterances, five short utterances and five longer ones and recorded with seven kinds of emotions: happiness, neutral, boredom, disgust, fear, sadness and anger [64]. The sentences are chosen to be semantically neutral and hence can be readily interpreted in all of the seven emotions simulated. Speech is recorded with 16 bit precision and 48 kHz sampling rate (later down-sampled to 16 kHz) in an anechoic chamber. The raw database contains approximately 800 sentences (7 emotions × 10 sentences × 10 actors + some second versions) [10]. The speech files are further evaluated by a subjective perception test with 20 listeners to guarantee the recognizability and naturalness of the emotions. Only utterances scoring higher than 80 % emotion recognition rate and considered natural by more than 60 % listeners are retained. The final numbers of utterances for the seven emotion categories in the Berlin database are: anger (127), boredom (81), disgust (46), fear (69), joy (71), neutral (79) and sadness (62) [10].The database comprises approximately 30 min of speech. 535 utterances are retained after evaluation by human subjects. 233 of 535 are uttered by male speakers, whereas the remaining 302 are uttered by female speakers. The length of the speech samples varies from 2 to 8 s.

5.4   Spanish Emotional Speech Database

As a part of the IST project Interface ("Multimodal Analysis/Synthesis System for Human Interaction to Virtual and Augmented environments"), an emotional speech database for Slovenian, English, Spanish, and French language has been recorded [22]. Among these four interface emotional speech databases Spanish emotional speech database is used for our work. Spanish database contains 184 sentences for each emotion which include numbers, words, sentences etc. The corpus comprises of recordings from two professional actors, one male and one female. Among 184 files, 1100 are Affirmative sentences, 101,134 are Interrogative sentences, 135,150 Paragraphs, 151,160 Digits, 161,184 Isolated words [35]. The Sentences of the database are selected from large collection of text which should contain emotionally neutral content. The emotional categories of Spanish database are anger, sadness, joy, fear, disgust, surprise and neutral. Subjective evolution tests are also made for Spanish database [22] which include 16 non-professional listeners(engineering students from UPC). Fifty six utterances were played (Seven per emotion long and short ones) Spanish emotional speech database has been recorded with AKG 320 condenser microphone in a silent room. Each speaker has assigned two sessions that were 2 weeks apart. The duration of each session is about 4 h for reading the script in all emotional states. Spanish Interface database contains 5520 sentences [22].

5.5   Real Time Emotional Database of a Driver

Before collecting the database recorded from the driver first one should identify the voice of the driver because it should be mixed with the co-passengers voice as well as the entertainment system and noise around the environment. In real time driving situation, the entertainment systems and the co-passengers decreases the performance of the emotion recognition system.

The Speaker recognition is generally divided into speaker identification and speaker verification. Speaker identification means based on the voice which one of the known group voice matches with the given voice. Speaker verification means person recognition based on the input voice. The voice based person identification plays a major role in speaker recognition applications. Each speaker has its own characteristics manner of speaking.

The speaker recognition technology uses the acoustic features of the speech to recognize the individual. The acoustic pattern reflect the anatomy which means means information about vocal card and behavioral properties like pitch, energy, etc. In practical driver identification based on acoustic features, the noise due to vehicle, environment and other co passenger's voice degrades the performance of the driver identification. In order to improve the robustness of the speaker identification, one has to filter out the speaker/driver voice from the mixture of voices.

Chapter 6
Classification Models

6.1 Introduction

Human is the best pattern recognizer. The human has a skill of identifying and recognizing the person from the thousands of people even after so many years with different aging, different light conditions, viewing conditions and with varying face expressions. This excited many researchers to focus on the pattern recognition systems to develop and make the machine as intelligent as the human. A set of measurements and observations about an object is a feature. Feature vector is a collection of such n features of the object ordered into n dimensional column vector. Class is the category to which object belongs to (grouped patterns which belongs to the same category). For given consideration, a collection of features with correct class information of an object is called a pattern. Examples of the patterns are face image, speech signal, bar code, fingerprint image, a word etc. Observing and extracting multiple features is a less complex task (in some cases it is obvious) for humans whereas for machines it is much more complex [33]. The pattern recognition is a science of Observing and extracting the patterns by machines. A pattern recognition system is an automatic system to classify the input pattern into a specific class. This system has two steps. First step is analysis. This step extract the feature from the pattern and second step is classification. This step recognize the objects based on the features. For classification step, the training set is needed. The training strategies are supervised learning and unsupervised learning [33, 48]. For supervised learning prior class information is needed. In unsupervised learning no class information is needed.

6.2 General Classification of Classification Models

The pattern recognition systems are commonly classified into four major methodologies.

- Statistical approach
- Syntactic or structural approach

© The Author(s) - SpringerBriefs 2015
K. R. Anne et al., *Acoustic Modeling for Emotion Recognition,*
SpringerBriefs in Electrical and Computer Engineering, DOI 10.1007/978-3-319-15530-2_6

- Template matching
- Neural networks

6.2.1 Statistical Approach

Statistical approach is based on the statistical properties and probabilities. patterns are represented as a set of features. One pattern has many features and each feature is converted to number and placed into a vector [26]. Each feature set is represented as a point in specific multidimensional space. Suppose, we have n features and is represented as a point in n dimensional vector space. To classify the patterns, the statistical approach measures the distance between the points in the multidimensional space. This approach establish a boundary to separate the different class patterns. The effectiveness depends on the how good the different class patterns are separated.

6.2.2 Syntactic or Structural Approach

The syntactic approach based on the complex relation between the features in a pattern. In this approach, pattern is represented as a hierarchical structure composed of substructures [33, 48]. All patterns related to one class contains same structural properties. That means the complex pattern is divided into small subpattern and each smallest subpattern is called primitive (also called codeword). So in this approach, shape is represented as a set of predefined primitives called code book.

6.2.3 Template Matching

Template matching approach is one of the simplest and widely used approach in pattern recognition to identify shapes in the image. In template matching, we should know the template of the pattern to be recognized. This template is obtained from the training set. The test pattern to be recognized is the match of the stored template with each possible position (in the image), each possible rotation, or each other geometric transformation of the template and compare each neighboring pixels to this template [33]. This approach calculates the correlation or similarity between the two points, curves and shapes of the same type.

6.2.4 Neural Networks

Neural network approach is parallel computing system consists of a structure with large number of processors and many interconnections between them.The interconnection sends the signal between one processor to other. The neural network trained itself and solves the complex problems based on the knowledge available [51]. Neural networks are inspire by the physiological information of the human brain and structure looks like the brain nervous system.

6.3   Selected Classification Models

The emotion recognition needs a database with a significant number of variables. This means a high dimensionality database is required. This high dimensionality database contains more similar features. In such situations, we need to reduce the dimensionality by only selecting the non-correlated features (information loss is very less) from the database. If we select the inadequate features, the accuracy will be reduced. So we need to acquire full knowledge about similarities and differences in the given data. Statistics are based on analyzing the high amount of data in terms of relationship between the variables in the data set. Each technique contains train database and test database. The train database contains different speech samples of different emotions. Test database contains one speech sample of unknown emotion. Each technique calculates the basis vector by using some statistical properties. The Statistical approach based pattern recognition techniques are discussed in this book. This approach is simplest and most widely used in practice for high dimensional data. The structural approach is very complicated for high data. The template matching techniques need high processing time to calculate the template. In the case of neural networks, the conventional concepts (number of layers,number of nodes,learning rate etc.) are not robust enough and depending on the complexity of the task. We cannot always easily formulate a mathematical definition of the problem to be solved.

These algorithms are widely used in several applications like signal processing and image compression. LDA and RDA uses the covariance or correlation matrix of given data.The LDA and RDA are used for multidimensional data set and are used as the dimensionality reduction (without any information loss) techniques.

The basis vector is of high dimensional, so we need to reduce the dimensionality. After forming the basis vector, the feature vector is calculated by projecting the train database speech samples into the basis vector. Then the matching is done by using distance measures. In this book, two data sets are taken. The first set is for training and the second set is for test.The training set contains many folders depending upon the selection of the data base. Each folder contains different emotional speech samples. Test set contains one speech sample. For this,the test speech samples are in different emotions.For different pattern recognition tasks, the statistical approach performs well for both supervised and unsupervised types. The selected pattern recognition algorithms discussed in this book are

- Linear Discriminant Analysis
- Regularized Discriminant Analysis
- Support Vector Machine
- K-Nearest Neighbor

6.3.1   Linear Discriminant Analysis

The classical Linear Discriminant Analysis (LDA) is also called as Fisher's Linear Discriminant (FLD). This method was developed by Robert Fisher in 1936. In this method, training and test sets are projected into the same subspace and the similarities between these data sets are identified. The Fisher's linear discriminant algorithm is explained in Fig. 3.1. LDA is appearance based linear subspace technique. So it uses statistics like mean and covariance. The calculation of the mean and co-variance is performed by using the train data set to form the data matrix X. The calculations of the mean and covariance are as shown in Eqs. 6.2 and 6.3

Calculation of mean and covariance matrix
First take the speech sample in the training set and convert this into a column vector. Doing this so, this convert all speech samples in the train set into column vectors. For example training set contains N speech samples and size of each speech sample is $M \times 1$. After converting all speech speech samples into column vectors,then append all columns. It forms a matrix called data matrix X with size $M \times N$. After forming data matrix,most used statistical measures (mean and covariance) have been calculated.

Calculate the Mean Speech Sample
The mean of an speech sample denotes the central location of the whole data and it is not necessarily same as the middle values. The mean of the random vector x is calculated from the Eq. 6.1

$$m = E[x] \tag{6.1}$$

$E[.]$ are the expected values of the argument x. Where x is the random sample corresponding column vector in the data matrix. Here columns of the data matrix $x(i)$ use the expression as shown in Eq. 6.2

$$m = \frac{1}{N} \sum_{i=1}^{N} x_i \tag{6.2}$$

where N is the number of speech samples in the training set, m represents the mean vector from the data matrix. It also represents the mean speech sample (when it is converted from column vector to matrix) in the training set.

Calculate the Covariance Matrix
Covariance measures the linear relationship between the two variables. So the dimension of the covariance is two. If we have a data set with more than 2 dimensions, then we have so many different covariance values. If we have n dimensional data set, then we have different covariance values. For example we have 3 dimensional data set with dimensions x, y, z. We calculate the covariance of x, y and the covariance of y, z and the covariance of x, z. The covariance matrix C is a matrix containing each entry as a covariance value. High covariance value indicates the high redundancy and

low value indicates low redundancy.The covariance matrix C of the random vector x is calculated using the Eq. 6.3 or 6.4

$$C = E[(x - m)(x - m)^T] \tag{6.3}$$

$$C = \frac{1}{N} \sum_{i=1}^{N} (x_i - m)(x_i - m)^T \tag{6.4}$$

$$C = AA^T \tag{6.5}$$

If we calculate the covariance matrix by using given Eq. 6.5 it takes high memory because of the dimensions of C. The size of A is $M \times N$. The size of C is $M \times M$ which is very large. It is not practical to calculate C as shown in Eq. 6.6. Let us consider the matrix L instead of C

$$L = A^T A \tag{6.6}$$

The dimension of L is $N \times N$ which is much smaller than the dimensions of C.

We could calculate eigenvalues and eigenvectors of covariance matrix. These eigenvectors and eigenvalues give the important information regarding the data. Eigenvectors give the uncorrelated variables. These uncorrelated variables are called principal components. The first principal component describes the high amount of variation. The eigenvalues of the covariance matrix describes the variance of the corresponding principal component i.e. it describes that the first principal component exhibits the highest amount of variation and the second principal component exhibits the second highest amount of variation and so on. Almost all appearance based techniques use statistical properties like mean and covariance to analyze the speech sample.Take the highest significant eigen vectors (principal components) and neglect the less significant eigenvectors.

Formation of a Feature Vector

After calculating eigenvectors of the covariance matrix, the dimensionality reduction takes place. Here we do not consider all the eigenvectors as principal components. We arrange all eigenvalues in the descending order and we take first few highest eigenvalues and corresponding eigenvectors. These eigenvectors e1, e2 and so on are the principal components as shown in Eq. 6.7

$$W = [e_1, e_2, ..., e_n] \tag{6.7}$$

neglect or ignore the remaining less significant eigenvalues and corresponding eigen vectors. These neglected eigenvalues represent a very small information loss.The principal component axis pass through the mean values. With these principal components (eigenvectors) we form a matrix called feature vector (also called eigen space). A new transformation matrix W is obtained by projecting the principal component on to the original data set. Then the data set is formed with new representation called feature space [70].

Derivation of a New Data Set

To derive a new data set with reduced dimensionality, take the transpose of the feature vector matrix (now each row of the matrix represents the eigenvector) and project this matrix on to the original data set with subtracted mean. This form the new transformation using the linear transformation of the original space into new reduced dimensional feature space by using Eq. 6.8

$$Y = W^T X \tag{6.8}$$

The sample mean of the data matrix X is m. In FLD, we have to calculate the mean of each class sample which is represented as m_i and i represent the specific class.

$$S_w = \frac{1}{N} \sum_{i=1}^{c} N_i S_i \tag{6.9}$$

Within class scatter matrix is given by Eq. 6.9. S_w calculates the amount of variance between the samples in each class. N represents the sample vectors x_i with n dimensionality and S_i is the sum of the covariance matrix of the samples in each class. S_i is calculated by using the Eq. 6.10

$$S_i = \frac{1}{N_i} \sum_{x}^{c} N_i S_i \tag{6.10}$$

S_i represents the class dependent scatter matrix. X_i represents the data matrix corresponding to class i. N_i represents the sample vectors present in class i. c represents the total number of classes. The between class scatter matrix is given by the Eq. 6.42

$$S_b = \sum_{i=1}^{c} (m_i - m)(m_i - m)^T \tag{6.11}$$

where

$$m = \frac{1}{N} \sum_{i=1}^{N} x_i \tag{6.12}$$

The overall or mixing scattering matrix is calculated by the covariance matrix of all speech samples as shown in Eq. 6.13

$$S_m = E[(X - m)(X - m)^T] = S_w + S_b \tag{6.13}$$

If S_w is nonsingular, we should solve transformation matrix W of generalized eigen problem. W should maximize the between class scatter matrix and minimize the within class scatter matrix.

$$S_b W = S_w W \lambda \tag{6.14}$$

If the transform $W = [w_1, w_2, ..., w_m]$ is applied, the average within-class variation will be

$$\tilde{S}_w = \frac{1}{N} \sum_{i=1}^{c} \left\{ \frac{1}{N_i} \sum_{i=1}^{c} \left(W^T x_i - \frac{1}{N_i} \sum_{i=1}^{c} (W^T x_i) \right) \left(W^T x_i - \frac{1}{N_i} \sum_{i=1}^{c} (W^T x_i) \right)^T \right\}$$

(6.15)

$$= w^T \left\{ \frac{1}{N} \sum_{i=1}^{c} c N_i S_i \right\} W$$

(6.16)

$$= w^T S_w W$$

(6.17)

Similarly, the average between-class variation will be

$$\tilde{S}_b = w^T S_b W$$

(6.18)

There are many solutions to solve the generalized eigen problem. After solving this problem, the result is

$$J(W) = \frac{\tilde{S}_b}{\tilde{S}_w} = \frac{|w^T S_b W|}{|w^T S_w W|}$$

(6.19)

which is called as Fisher's criterion [28]

$$\tilde{W} = \underset{W}{argmax} J(W) = \underset{W}{argmax} \frac{|\tilde{S}_b|}{|\tilde{S}_w|} = \underset{W}{argmax} |w^T S_b W| |w^T S_w W|$$

(6.20)

Or, for each column vector W_i of W.

The quadratic form has optimal solution $\lambda_i = \frac{w_i^T S_b W_i}{w_i^T S_w W_i}$

By applying derivatives on both sides and solving we get

$$S_b W_i - S_w W_i \lambda_i = 0 \Rightarrow S_b W_i = S_w W_i \lambda_i$$

(6.21)

One method for solving this eigen problem is to take the inverse of S_w and solve the following problem by using matrix $S_w^{-1} S_b$. λ is a diagonal matrix containing the eigenvalues of matrix $S_w^{-1} S_b$

$$S_w^{-1} S_b W = W \lambda$$

(6.22)

For W we must calculate the eigenvalues values and eigenvectors by using the singular value decomposition of $S_w^{-1} S_b$. This algorithm is optimal only when the scatter matrix is non singular. If S_w is singular then we get a warning that matrix is close to singular or badly scaled. This is a singularity problem and occurs due to high dimensional and low sample size speech data.

In general there exists three methods to solve the singularity problem in LDA.

Subspace method deal with the singularity problem. In this method first apply PCA, an intermediate dimensionality reduction step, to reduce the data dimensionality before LDA is applied. The algorithm is known as PCA + LDA, or subspace LDA. In this two-stage PCA + LDA algorithm, the discriminant stage is preceded by a dimensionality reduction stage using PCA. The dimensionality of the subspace transformed by PCA is chosen such that the reduced total scatter matrix in this subspace is nonsingular, so that classical LDA can be applied.

Regularization techniques can also be applied to deal with the singularity problem of LDA. The algorithm is known as Regularized Discriminant Analysis, or RDA in short. The key idea is to add a constant to the diagonal elements of S_w. The singularity problem elimination depends on choosing the value of the regularized parameter. In this book we concentrate more on this method in order to eliminate the singularity problem.

The null space LDA (NLDA) was also proposed to overcome the singularity problem, where the between-class distance is maximized in the null space of the within-class scatter matrix. The singularity problem is thus avoided implicitly. By first removing the null space of the total scatter matrix the efficiency of the algorithm can be improved. It is based on the observation that the null space of the total scatter matrix is the intersection of the null spaces of the between-class and within-class scatter matrices.

The procedure for solving the singularity problem, by using the Regularized Discriminant Analysis is discussed in next section.

6.3.2   Regularized Discriminant Analysis

The block diagram for both Linear and Regularized discriminant Analysis are given in the Fig. 6.1. If S_w is non singular it comes under Linear Discriminant Analysis else it comes under Regularized Discriminant Analysis.

The key idea behind the regularized discriminant analysis is to add a constant to the diagonal elements of with in class scatter matrix S_w of the speech samples is shown in Eq. 6.23.

$$S_w = S_w + \lambda I \tag{6.23}$$

where λ is regularized parameter which is relatively small such that S_w is positive definite. In our paper the value of λ is 0.001. It is somehow difficult in estimation of regularization parameter value in RDA as higher values of λ will disturb the information in the within class scatter matrix and lower values of λ will not solve the singularity problem LDA [70, 77].

The transformation matrix W is calculated by using $S_w^{-1} S_b W = W\lambda$. Once the transformation matrix W is given, the speech samples are projected on to this W. After projection, the Euclidian distance between each train speech sample and the test speech sample are calculated, the minimum value among them will classify the result [35].

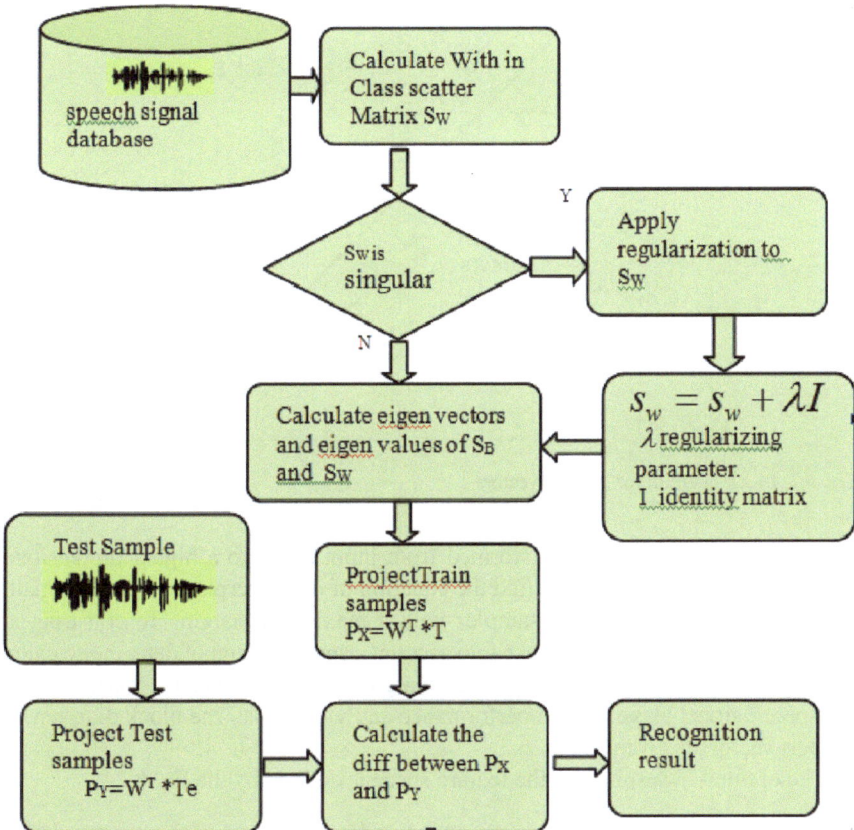

Fig. 6.1 Block diagram for Regularized Discriminant Analysis

6.3.3   Support Vector Machine

Support vector machine is an effective approach used for binary classification. This can be extended for multi class classification using a combination of binary class support vector machines. The basics of the SVM can be found in references [8, 13, 23, 30, 71]. There are many ways for multi category SVM in the literature among them most popular ones are one-against-all and one-against-one. Most of the people are used one against all approach for classification of multi class problem.

A set of features like prosody and spectral are extracted from the speech signal and are used as input for training the SVM classifier. MFCC, Pitch and intensity are the extracted prosody and spectral features from the speech signals of different emotional classes. All these values are grouped in the form of a matrix. Each row of the matrix is a representation of speech signal and the column vectors are corresponding to values of different features.

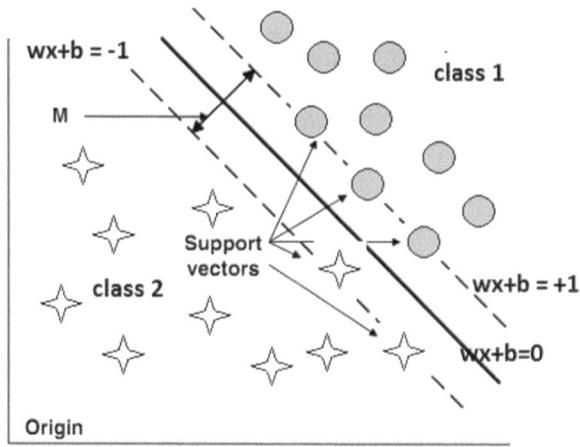

Fig. 6.2 Block diagram for support vectors

The feature vectors are transformed from input space to a higher dimensional feature space. Now we have to find the equation of the hyperplane which optimally separates the training speech samples with a maximum margin. To construct an optimal hyperplane one has to take into account a small amount of data, the so called support vectors, which determines this margin [42]. As higher the margin, more the speech signal classification performance and vice versa. The block diagram for support vectors is given in Fig. 6.2

The optimal hyperplane in the feature space is defined by the Eq. 6.24

$$(w.x) + b = 0 \qquad (6.24)$$

Where x is our row vector of corresponding speech sample, w is weight vector and b is the bias. The distance from the hyperplane to the closest points of the two classes on both sides of the hyperplane is called margin M of the hyperplane. To maximize the margin M, has to be minimized subject to conditions.

$$\min_{w,b} \frac{1}{2}\|w\|^2 \, subject \, to \, y_i.((w.x_i) + b) \geq 1, i = 1, 2, ...n. \qquad (6.25)$$

This is a quadratic optimization problem. Lagrangian function is used to solve this problem and obtain appropriate Langrange multipliers ($\alpha \alpha_k$).

$$L(w, b, \alpha) = \frac{1}{2}\|w\|^2 - \sum_{i=1}^{n} \alpha_i(y_i.((w.x_i) + b) - 1) \qquad (6.26)$$

The Lagrangian L has to be minimized with respect to the primal variables w and b and maximized with respect to the dual variables Then we obtain weight vector for the optimal hyperplane and is a linear combination of support vectors.

$$w = \sum_{i=1}^{n} \alpha_i y_i x_i \qquad (6.27)$$

Here $y_i x_i$ are called support vectors, y_i represents class lable and x_i represents the training sample. Here we trained four classes of emotions(Happy, Neutral, Anger and Sad) so we obtain four models one for each emotion. In each model we have support vectors $y_i x_i$, α_i and bias b values. By using this support vectors now we can classify our test speech sample by using the decision function

$$f(x) = sign(\sum_{i=1}^{n} \alpha_i y_i x_i . x + b) \tag{6.28}$$

The dot product is applied between each test speech sample with the support vectors, alpha and bias values obtained during the training phase. Because the data is transformed from input space to high dimensional feature space, with the mapping function $\Phi : R^n \rightarrow H$ After obtaining support vectors a kernel function is applied here we used linear kernel

$$f(x) = sign \left(\sum_{i=1}^{n} \alpha_i y_i K(x_i, x) + b \right) \tag{6.29}$$

where $K(x_i, x) = x_i . x_j$ is a linear kernel. As a result we obtain four models, one for each emotional class in training phase. Basing on these models and the generated support vectors we will classify the emotion of the test speech sample.

6.3.4 K-Nearest Neighbor

KNN is a non-parametric method for classifying speech samples based on closest training samples in the feature space [54]. Similar to SVM and RDA the speech samples are given as input to the KNN classifier. Nearest Neighbor classification uses training samples directly rather than that of a model derived from those samples. It represents each speech sample in a d-dimensional space where d is the number of features. The similarity of the test samples with the training samples is compared using Euclidian distance. Once the nearest neighbor speech samples list is obtained the input speech sample is classified based on the majority class of nearest neighbors.

$$y' = arg\ max_v \sum_{(x_i, y_i) \in D_z} I(v = y_i) \tag{6.30}$$

where

v	class label
y_i	class label for one of the nearest neighbors
I(.)	function returns 1 if its arg is T , 0 otherwise be the test sample
(x_i, y_i)	is the test speech sample

Every neighbour has the same impact on the classification. One way to reduce the impact is to weight the neighbour according to its distance.

$$w_i = \frac{1}{d(y', y_i)^2} \tag{6.31}$$

Training examples located far away are the weaker impact on the classification. So classification can be done based on the following equation.

$$y' = arg\ max_v \sum_{(x_i, y_i) \in D_z} w_i * I(v = y_i) \tag{6.32}$$

6.4   Distance Measures

We have several linear distance measuring techniques. In [36], Moon and Phillips explains the different distance measure. They are euclidean distance, mahalanobis distance, Minkowski distance, city block and cosine distance metric. Let us take a training set of N sample files. Then calculate the feature vector Y with these sample files i.e. Y has N number of $(K \times 1)$ column vectors as y1,y2,...yN. The feature vector of test sample is y_{tst}. Calculate the distance d between y_i and y_{tst} by using various distance measures. Where i represents the ith column vector.

$$y_{(i=1,2,...,k)} = \begin{pmatrix} y_{11} & y_{12} & \cdot & \cdot & y_{1N} \\ y_{21} & \cdot & \cdot & \cdot & y_{2N} \\ \cdot & \cdot & \cdot & \cdot & \cdot \\ \cdot & \cdot & \cdot & \cdot & \cdot \\ y_{k1} & \cdot & \cdot & \cdot & y_{kN} \end{pmatrix} \tag{6.33}$$

6.4.1   Euclidian Distance

The euclidean distance is a straight line distance and which is commonly used distance measure in many applications. This distance gives the shortest distance between the two sample files or vectors it is same as the Pythagoras equation in 2 dimensions [49]. It is sum of squared distance of two feature vectors $(y - tst, y_i)$ is given by Eq. 6.39

$$d^2 = (y_{tst} - y_i)^T (y_{tst} - y_i) \tag{6.34}$$

$$d = \sum_{j=1}^{n} |y_{tstj} - y_{ij}| \tag{6.35}$$

The Euclidean distance is sensitive to both adding and multiplying the vector with some factor or value.

6.4.2 Standardized Euclidian Distance

The standardized euclidean distance is defined as shown in Eq. 6.36

$$d^2 = (y_{tst} - y_i)^T D^{-1}(y_{tst} - y_i) \tag{6.36}$$

For this distance measure consider the variance v_j^2 of the jth element of y whose observations are the $y_i's$. (i = 1, 2,..,N). D is the diagonal matrix with diagonal elements contains the variance v_j^2 The standardized euclidean distance measure does not improves the detection rate as compared with the euclidean distance measure. If we use the standardized euclidean distance measure, the number of eigenvectors increases then detection rate is decreasing.

6.4.3 Mahalanobis Distance

Mahalanobis distance comes from the Gaussian multivariate probability density function as defined in Eq. 6.37

$$p(x) = (2\pi)^{-d/2}|C|^{-1/2}exp(-1/2(x - m)^T C^{-1}(x - m)) \tag{6.37}$$

where $(x - m)^T C^{-1}(x - m)$ is called squared mahalanobis distance, which is very important in characterizing the distribution. Where C is the estimated covariance matrix of y whose observations are the $y_i's$ s. The mahalanobis distance of two feature vectors y_{tst} and y_i is calculated using the Eq. 6.38.

$$d^2 = (y_{tst} - y_i)^T D^{-1}(y_{tst} - y_i) \tag{6.38}$$

The Mahalanobis distance and the standardized euclidean distance measure gives almost the same performance

6.4.4 Cityblock Distance

The city block distance is defined by using the Eq. 3.33.

$$d = \sum_{j=1}^{n}(y_{tstj} - y_{ij}) \tag{6.39}$$

6.4.5   Minkowski Distance

$$d = \left(\sum_{j=1}^{n} |(y_{tstj} - y_{ij})|^p \right)^{1} / p \tag{6.40}$$

The city block distance is a special case of the minkowski distance measure when p=1 and the euclidean distance measure is special case of the minkowski distance measure when p=2.

6.4.6   Cosine Distance

The cosine distance is defined by using the Eq. 3.35.

$$d = (1 - y_{tst}^T y_i / (y_{tst}^T y_{tst})^{1/2} (y_i^1 y_i)^{1/2}) \tag{6.41}$$

The cosine distance measure and the euclidean distance gives almost same accuracy [41]. When compared to all given distance measures, euclidean distance measure gives the better or almost same performance. So Euclidean distance is used in most of the applications. The mahalanobis distance performs poorly when compared to Euclidean distance. Wendy et al claims that the mahalanobis distance measure proposed by Moon's [36] gives better performance when compared to the mahalanobis distance measure, which is defined as in Eq. 6.42

$$d = - \sum_{j=1}^{n} \sqrt{\frac{\lambda_j}{\lambda_j + \alpha^2}} y_{tst,j} y_{i,j} \tag{6.42}$$

Take $\alpha = 0.25$

$$\sqrt{\frac{\lambda_j}{\lambda_j + \alpha^2}} \approx \frac{1}{\sqrt{\lambda_j}} \tag{6.43}$$

Then distance measure formula is changed as

$$d = - \sum_{j=1}^{n} 1 \sqrt{\lambda_j} y_{tst,j} y_{i,j} \tag{6.44}$$

Where λ_j is the jth eigenvalues corresponding to j^{th} eigenvector.

6.4.7 Hausdorff Distance

In this section we used a special distance metric which is able to compute the distance between different sized matrices having a single common dimension, like the acoustic matrices representing our speech feature vectors. It derives from the Hausdorff metric for sets [6, 16].

The Hausdorff distance measures the extent to which each point of a 'model' set lies near some point of an sample set and vice versa. Unlike most vector comparison methods, the Hausdorff distance is not based on finding corresponding mode and sample points [67]. Thus, it is more tolerant of perturbations in the location of points because it measures proximity rather than exact superposition.

However, the Hausdorff distance is extremely sensitive to outliers. If two sets of points $A(a1, a2)$ and $B(b1, b2, b3)$ are similar, all the points are perfectly superimposed except only one single point in A which is far from any point in B, then the Hausdorff distance is determined by that point and is large. This sensitivity to outliers is not acceptable. So some modified Hausdorff distances and directed Hausdorff distance are used for the voice recognition.

6.4.8 The Directed Hausdorff Distance

The distance between two points a and b is defined as $d(a, b) = \| a - b \|$. Here, we not only compute the distance between the point a in the finite point set A and the same value in the finite point set $B = b1, ...b_{Nb}$ but also compute the distances between the a_t and its two neighbor values b_{t-1} and b_{t+1} in the finite point set B, respectively, and then minimize these three distances [6].

$$d(a, B) = \min_{b \in B} d(a, b) = \min_{b \in B} \| a - b \| \qquad (6.45)$$

The directed Hausdorff metric h(A,B) between the two finite point set $A = a_1, ...a_{Nb}$ $B = b_1, ...b_{Nb}$ is defined as follows

$$h(A, B) = \max_{a \in A} d(a, B) = \max_{a \in A} \min_{b \in B} d(a, b) \qquad (6.46)$$

$$h(A, B) = \left\{ \max_{a \in A} \left\{ \min_{b \in B} \| a - b \| \right\} \right\} \qquad (6.47)$$

In this way, h(B,A) is

$$h(B, A) = \max_{b \in B} d(b, A) = \max_{b \in B} \min_{a \in A} d(b, a) \qquad (6.48)$$

$$h(B, A) = \left\{ \max_{b \in B} \left\{ \min_{a \in A} \| b - a \| \right\} \right\} \qquad (6.49)$$

where d is any proper metric between the points of sets A and B (for example, the Euclidean distance). It is termed also as forward Hausdorff distance, while h(B,A) represents the backward Hausdorff distance for sets. Thus, we obtain the general definition for the Hausdorff distance for sets as follows [6, 16]

$$h(A, B) = \max\{h(A, B), h(B, A)\} \tag{6.50}$$

from the Eqs. 6.47, 6.49, 6.51 the next Hausdorff distance formula is obtained:

$$h(A, B) = \max\left\{\sup_{a\in A}\inf_{b\in B} d(a, b), \sup_{b\in B}\inf_{a\in A} d(a, b)\right\} \tag{6.51}$$

The components of this Hausdorff distance, $\sup_{a\in A}\inf_{b\in B} d(a,b)$ and $\sup_{b\in B}\inf_{a\in A} d(a,b)$ are sometimes termed as forward and backward Hausdorff distances of A to B. Let us consider now matrices having a single common dimension (the number of rows), instead of sets. Thus $A = (a_{ij})_{n\times m}$ and $A = (b_{ij})_{n\times p}$. Let us introduce two more helping vectors or auxiliary vectors, $y = (y_i)_{p\times 1}$ and $z = (z_i)_{m\times 1}$, then compute $\| y \|_p = \max_{0\leq i\leq p} \| y_i \|$ and $\| z \|_m = \max_{0\leq i\leq m} \| z_i \|$, respectively. With these notations a new nonlinear metric d has been created having the following form:

$$d(A, B) = \max\left\{\sup_{\|y\|_p\leq 1}\inf_{\|z\|_m\leq 1} \| B_y - A_z \|, \sup_{\|z\|_m\leq 1}\inf_{\|y\|_p\leq 1} \| B_y - A_z \|\right\} \tag{6.52}$$

This restriction based metric represents the Hausdorff distance between the sets $B(y :\| y \|_p\leq 1)$ and $A(\| z \|_m\leq 1)$ in the metric space R^n. Therefore it can be written using the following equation:

$$d(A, B) = H(B(y :\| y \|_p\leq 1), A(\| z \|_m\leq 1) \tag{6.53}$$

From th Eq. 6.53 we have

$$B_y - A_z = \sum_{k=1}^{p} b_{ik}y_k - \sum_{j=1}^{m} a_{ij}z_j \tag{6.54}$$

$$\| B_y - A_z \|_n = \max_{1\leq i\leq n} \| \sum_{k=1}^{p} b_{ik}y_k - \sum_{j=1}^{m} a_{ij}z_j \| \tag{6.55}$$

Therefore, it results the following equation:

$$\sup_{\|y\|_p\leq 1}\inf_{\|z\|_m\leq 1} \| B_y - A_z \|_n = \sup_{\|y\|_p\leq 1}\inf_{\|z\|_m\leq 1}\max_{1\leq i\leq n} \| \sum_{k=1}^{p} b_{ik}y_k - \sum_{j=1}^{m} a_{ij}z_j \| \tag{6.56}$$

This can be seen as a max min optimization problem and according to the classical J.vonNeumann min max theorem [6] we get:

$$\sup_{\|y\|_p\leq 1}\inf_{\|z\|_m\leq 1} \| B_y - A_z \|_n = \inf_{\|z\|_m\leq 1}\sup_{\|y\|_p\leq 1} \| B_y - A_z \|_n \tag{6.57}$$

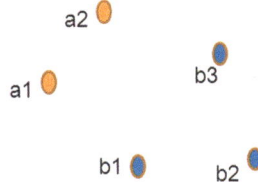

Fig. 6.3 Given two sets of points A and B, find h(A,B)

Next, after eliminating the terms y and z from the above formula, we finally obtain the following Hausdorff based distance

$$d(A, B) = \max\left\{ \sup_{1\leq k\leq p} \inf_{1\leq i\leq m} \sup_{1\leq i\leq n} |b_{ik} - a_{ij}|, \sup_{1\leq i\leq m} \inf_{1\leq k\leq p} \sup_{1\leq i\leq n} |b_{ik} - a_{ij}| \right\}$$
(6.58)

The resulted nonlinear function d verifies main distance properties:

- Non-negativity: $d(A, B) \geq 0$
- Symmetry: $d(A, B) = d(B, A)$
- Triangle inequality: $d(A, B) + d(B, C) \geq d(A, C)$

While not representing a Hausdorff metric anymore, the Hausdorff-based distance d given by Eq. 6.58 constitutes a satisfactory discriminator between the vocal feature vectors [27]. It defines a new metric topology on the space of all matrices $\{A\}$, that is not equivalent but comparable with that induced by Hausdorff topology. The newly introduced distance will be successfully used in the voice feature vector classification process.

Example

Finding the distance between a_1 and $b'_j s$, a_2 and $b'_j s$ and keep the shortest distances. Finally find the larger distance among the two shortest distances a_1 and $b'_j s$,a_2 and $b'_j s$

From the Fig. 6.3 compute the distance between a_1 and $b'_j s$ and keep the shortest distance, similarly a_2 and $b'_j s$ and keep the shortest distance. Finally find the largest of the two distances from the Figs. 6.4, 6.5. Now we can say that any point of A is at most distance $h(A, B) = d(a1, b1)$ to some point of B.

6.5 Summary

This chapter gives the pattern recognition basics and explains different existing pattern recognition approaches. Those are statistical approach, syntactic approach, template matching and neural network approaches. The statistical approach based techniques in comparison with other approaches are discussed in theoretical perspective. Different distance measure techniques are defined and compared. In distance

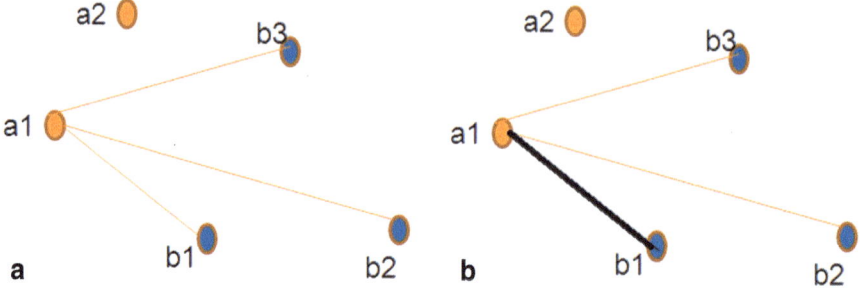

Fig. 6.4 Compute the distance between a_1 and $b'_j s$ (b) shortest distance a_1 to b_1

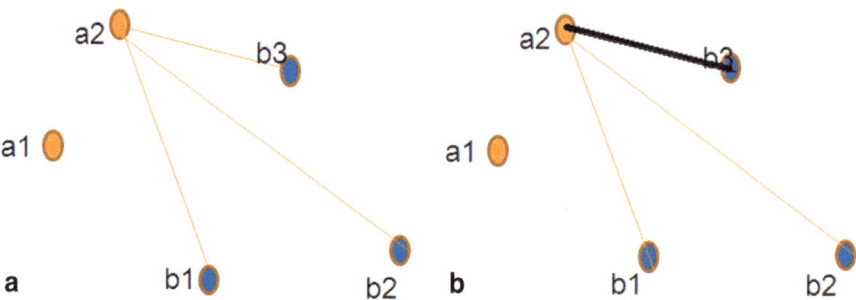

Fig. 6.5 Compute the distance between a_2 and $b'_j s$ (b) shortest distance a_2 to b_3

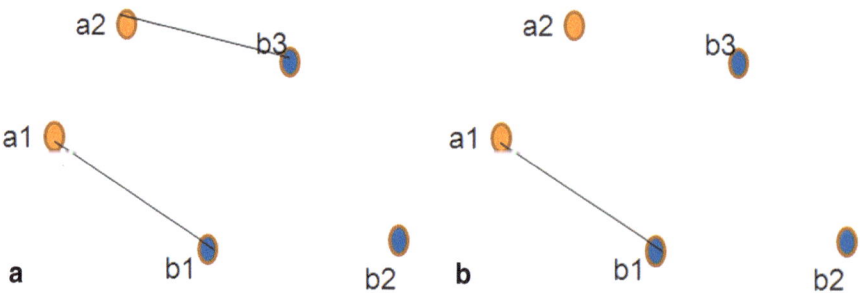

Fig. 6.6 Find the largest of the two distances from a_1 and a_2 (b) largest distance among those two shortest distance

measures,Euclidean distance gives the high performance but it is sensitive to both adding and multiplying the vector with some factor or value. In this thesis work, we used Euclidean distance, which is able to compute the distance between different matrices.

Chapter 7
Comparative Analysis of Classifiers in Emotion Recognition

7.1 Introduction

In this work the emotion recognition performance is determined by each classifier LDA, RDA, SVM and KNN for each database. A comparative analysis is done elaborately for all these classifiers as well as for databases.

7.2 Emotions Used in This Work

In order to validate the results of different classifiers on happy, neutral, anger and sad emotional classes, recognition tests were carried out in two phases like baseline results and feature fusion results.

In the first phase all the classifiers have been trained using prosody and spectral features of Berlin and Spanish emotional speech samples and the results are shown in Table 7.1. It seems that spectral features provide higher recognition accuracies than prosody ones. According to these results, it is observed that even though prosody parameters show very low class separability it does not perform so badly. It shows an accuracy in the range of 40– 67 % for both the databases. Results with spectral statistics seems rather modest reaching an accuracy of 50–70 % for all the classifiers, except for RDA which reaches 78.75 %. For further improvement in the performance of all the classifiers, feature fusion technique is used in the next phase.

7.3 Analysis of Results with Each Emotion for Each Classifier

Feature fusion is done by combining both prosody and spectral features. By doing this, the emotion recognition performance of the classifiers is effectively improved when compared with baseline results and are shown in Table 7.2. Among these RDA and SVM performs considerably better when compared with remaining classifiers.

© The Author(s) - SpringerBriefs 2015   55
K. R. Anne et al., *Acoustic Modeling for Emotion Recognition*,
SpringerBriefs in Electrical and Computer Engineering, DOI 10.1007/978-3-319-15530-2_7

Table 7.1 Emotion recognition percentage accuracy of various classifiers (LDA, RDA, SVM, and KNN) over Berlin and Spanish databases using prosody and spectral features

Classifier	Berlin		Spanish	
	Prosody (%)	Spectral (%)	Prosody (%)	Spectral (%)
LDA	42.0	51.0	40.0	49.0
RDA	67.0	78.75	44.0	61.0
SVM	55.0	68.75	60.25	64.5
kNN	57.0	60.7	58.25	67.75

Table 7.2 Emotion recognition percentage accuracy of various classifiers (LDA, RDA, SVM, and KNN) over Berlin and Spanish databases using feature fusion

Classifier	Feature Fusion	
	Berlin (%)	Spanish (%)
LDA	62	60
RDA	80.7	74
SVM	75.5	73
kNN	72	71.5

The overall recognition performances of these four classifiers for Berlin and Spanish databases is shown in Fig. 7.1. Horizontal axis represents the name of the classifier and vertical axis represents the accuracy rate. The efficiency of each classifier is improved by 20 % approximately for each classifier when compared with base line results.This proves that feature fusion is a best technique for improving the efficiency of the classifier.

7.4 Analysis of Results with Confusion Matrices

The confusion matrices of RDA classifier for both the databases is shown in the Table 7.3. By observing the results, we can say that the emotions happy is confused with anger and the emotion neutral is confused with sad. This occurs in all the features but the amount of variation is more by using individual features and is less by using feature fusion. The reason for this confusion is explained by using valence arousal space. Prosodic features are able to discriminating the emotions(happy, anger) from high arousal space to emotions(neutral, sad) from low arousal space, but there exists a confusion among the emotions in the same arousal state. By using spectral features and feature fusion the confusion among the emotions in the same arousal state is reduced.

The recognition accuracy of each emotion are examined separately with all the classifiers and are shown in Table 7.4. The left column of the Table 7.4 shows the classifiers and top row shows the emotion. Each cell represents the recognition accuracy of the emotion by the corresponding classifier. With Berlin database, the emotion recognition rate of the classifiers RDA and SVM are comparable with each

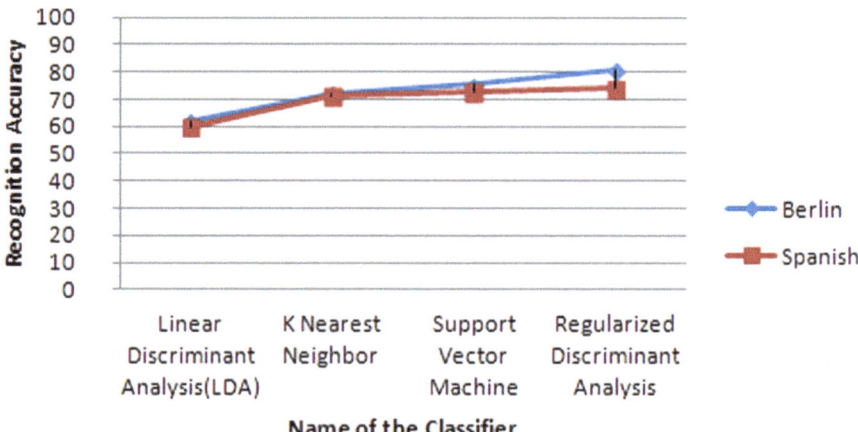

Fig. 7.1 Comparison of emotion recognition accuracy performance of different classifiers using Berlin database

Table 7.3 RDA based emotion classification performance in percentage using featurefusion technique (a) using Berlin database speech utterances (b) using Spanish database speech utterances

Berlin	Happy	Neutral	Anger	Sad
Happy	73	4	20	3
Neutral	7	70	–	21
Anger	14	–	83	3
Sad	–	3	–	97
		(a)		
Spanish	Happy	Neutral	Anger	Sad
Happy	71	5	15	9
Neutral	–	60	14	26
Anger	22	–	67	1
Sad	1	2	–	97
		(b)		

other for all the emotions. The emotion anger is identified more with kNN classifier for both databases.

The graphical representation of efficiencies of these classifiers is shown in the Fig. 7.2. The Blue, Red, Green and violet bars or bars from front to back comprises the efficiencies of LDA, kNN, SVM and RDA respectively. Analysis of emotions happy, neutral, anger and sad are represented by the bars from left to right.

Table 7.4 Recognition accuracy percentage for emotions(happy, neutral, anger and sad) with various classifiers using both the databases (a) Berlin database (b) Spanish database

Berlin	Feature fusion			
Algorithm	Happy	Neutral	Anger	Sad
LDA	49	59	68	72
RDA	73	70	83	97
SVM	70	65	74	93
kNN	55	63	93	77
		(a)		
Spanish	Feature fusion			
Algorithm	Happy	Neutral	Anger	Sad
LDA	49	56	65	70
RDA	71	60	67	97
SVM	67	77	65	83
kNN	63	70	88	65
		(b)		

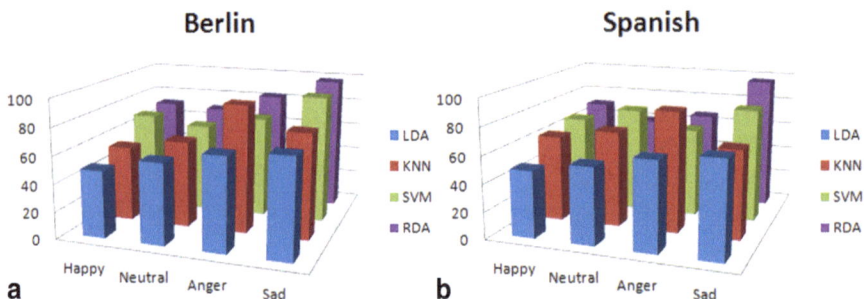

Fig. 7.2 Comparison of emotion recognition performances of different classifiers using **a** Berlin database **b** Spanish database

7.5   Brief Overview of ROC Curves

The shape of the ROC Curve and the area under the curve helps to estimate the discriminative power of a classifier. The area under the curve can have any value between 0 and 1 and it is a good indicator of the goodness of the test. From the literature survey the relationship between Area under curve and its diagnostic accuracy are shown in the Table 7.5. From the Table 7.5 it is observed that as the Area under curve (AUC) reaches 1.0, the performance of the classification technique is excellent in classifying emotions, if AUC is less than 0.6 the performance of classification technique is poor, if the AUC is in between 0.6 and 0.9 then the results obtained are satisfactory if it is less than 0.5 the classification technique is not useful.

Table 7.5 The relationship between area under curve and its diagnostic accuracy

Area under curve	Diagnostic accuracy
0.9 – 1.0	Excellent
0.8 – 0.9	Very good
0.7 – 0.8	Good
0.6 – 0.7	Sufficient
0.5 – 0.6	Bad
< 0.5	Test not useful

7.5.1 Analysis of Results with ROC Curves

To plot an ROC curve to our four emotional classes, we first divide speech samples into positive and negative emotions. Happy and Neutral speech samples are positive emotions and speech samples of Anger, Sad are negative emotions. Sensitivity evaluates how the test is good at detecting positive emotions and Specificity evaluates how good the negatives emotions are discarded from positive emotions. ROC curve is a graphical representation of the relationship between both sensitivity and specificity.

The Fig. 7.3. shows the performance comparison of different classifiers. The Table 7.6 shows the values extracted from ROC plot for different classification Algorithms over Spanish database. The results obtained using ROC curve are comparable with each other. The AUC is greater than 0.8 for RDA and SVM which means the diagnostic accuracy of these classifiers is very good, similarly the AUC is greater than 0.6 for kNN and LDA which means the diagnostic accuracy of these classifiers is sufficient from Table 7.5.

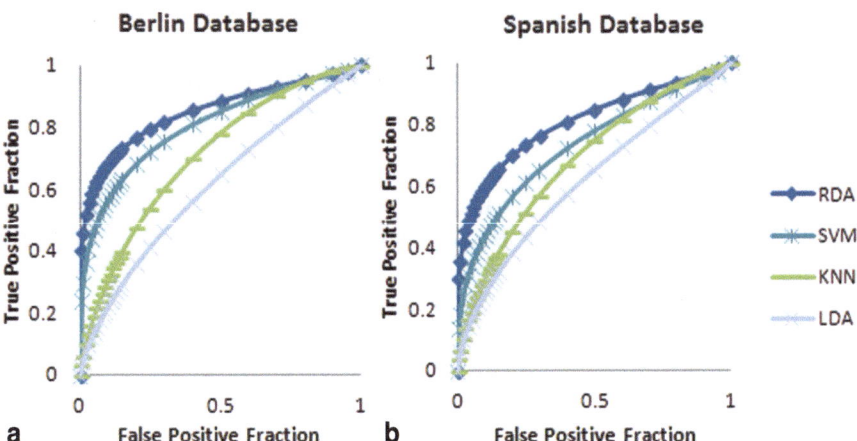

Fig. 7.3 Shows the comparison of performances of different classifiers using **a** Berlin database **b** Spanish database

Table 7.6 Shows the values (*Accu*: Accuracy, *Sens*: Sensitivity, Spec:Specificity, *AUC*: Area Under Curve)extracted from ROC plot for different classifiers for (a) Berlin database and (b) Spanish database

Algorithm	Accu	Sens	Spec	AUC
RDA	74.0%	70.7%	78.4%	0.814
SVM	72.0%	72.2%	71.8%	0.734
kNN	71.8%	75.1%	69.2%	0.688
LDA	62.0%	60.3%	64.3%	0.619
		(a)		
Algorithm	Accu	Sens	Spec	AUC
RDA	80.8%	75.7%	88.2%	0.854
SVM	75.5%	72.0%	80.4%	0.806
kNN	72.0%	67.5%	79.7%	0.709
LDA	60.0%	57.7%	60.8%	0.601
		(b)		

From the Table 7.6 it is observed that the AUC is above 0.8 for RDA and above 0.7 for SVM which means the diagnostic accuracy of the classifiers are very good and good respectively, in the same way the AUC is more than 0.6 for kNN and LDA which means the diagnostic accuracies of these classifiers is sufficient from Table 7.5.

7.6 Summary

The emotion recognition accuracy using Acoustic information is systematically evaluated by using Berlin and Spanish emotional databases. This has been implemented by a variety of classifiers including LDA, RDA, SVM and KNN using various prosody and spectral features. The emotion recognition performance of the classifiers is obtained effectively with feature fusion technique. An extensive comparative study has been made on these classifiers. The results of the evaluation have showed that RDA yields better recognition performance and SVM also gives good recogniiton performance compared with other classifiers. The use of feature fusion technique instead of using individual features enhances the recognition performance. The experimental results suggests that recognition accuracy is improved by 20% approximately for each classifier with feature fusion.

Chapter 8
Summary and Conclusions

This chapter summarizes the research work on speech emotion recognition presented in this book. Conclusions drawn from the present work, scope and future work.

This book organized into eight chapters. The first chapter introduces the basics of acoustic speech signals and different perspectives of emotion. The second chapter reviews different prosodic features and their importance in recognizing the emotion. The third chapter discusses different spectral features and their importance in recognizing emotions. The fourth chapter introduces different feature fusion techniques. The fifth chapter discusses different emotional speech corpora used for this work. The sixth chapter introduces different classification models. The seventh chapter discusses the comparative analysis of different classifiers in recognizing emotions. The eighth chapter concludes the present work and present some new ideas in this area for future research.

8.1 Summary of the Present Work

The aim of acoustic modeling for emotion recognition is to extract different acoustic features to recognize the emotion of a person. In this work we came to know that, fusion of different acoustic features like prosody and spectral will significantly increases the performance. Also in this work, we gave a weighage to features that gave more emotion specific information and omitted other features. The feature selection plays an important role in recognizing the emotion in order to increase the performance. Each and every feature is implemented by writing the code in matlab. In this work we emphasized on performing the evaluation over different emotional databases and check the robust ness of different classification techniques.

© The Author(s) - SpringerBriefs 2015
K. R. Anne et al., *Acoustic Modeling for Emotion Recognition,*
SpringerBriefs in Electrical and Computer Engineering, DOI 10.1007/978-3-319-15530-2_8

8.2   Conclusion from the Present Work

Prosodic features extracted from the speech signals give emotion specific information. Among various prosodic features proposed, Energy and Pitch perform better in recognizing emotions. But the drawback with these features is, even though they well differentiate the emotions in different valance state, they are unable to differentiate the emotions in the same arousal state. This drawback can be eliminated by using Mel Frequency Cepstral Coefficients, i.e spectral features outperform the prosodic features in recognizing the emotions. They well differentiate the emotions in the same arousal state. The performance is further improved by using feature fusion i.e by combining spectral and prosodic features. All this is done with different classifiers like Linear Discriminant Analysis (LDA), Regularized discriminant Analysis (RDA), Support vector machines (SVM) and k nearest neighbor (KNN). Among these because of singularity problem the LDA gives lowest performance and this can be eliminated by using regularization technique in RDA. The performance obtained with KNN is good and the better performance is given by both RDA and SVM.

8.3   Scope and Future Work

The work done in this book is by using Berlin and Spanish simulated emotional speech databases. This can be extended to emotions which are collected from real life environments naturally. The expressions of emotions are common in any language and are independent of speaker and gender also. So the work can be extended to cross-lingual emotion recognition. In this the speech samples are trained with one language and the test sample is from another language. The work in this book can be carried out using para linguistic aspects of speech but to improve the performance his can be extend with linguistic aspects of speech also. In real time applications this work can be applied to detect the emotion of the driver to alert him from an accident. Along with vocal expressions, facial expression are also important in determining the exact emotion of the driver. The emotion classification task with complete set of features is a time taking task so we have to select some features which best detects the emotions and leave the remaining features. This can be done with feature subset selection algorithms like Sequential forward selection and Sequential forward floating selection. By using these algorithms will increase the performance of the classifier as well as reduces the computation time. Now a days there exists new measures like Precision, Recall and F-Measure which are used to estimate the performance of the classifier. We can calculate these measures by using the values extracted from confusion matrices.

References

1. Noelia Alcaraz Meseguer, 2009, *Speech Analysis for Automatic Speech Recognition*, Master Thesis, Norwegian University of Science and Technology, Norway.
2. Andersson, Tobias. 2004. Audio classification and content description. Lulea University of Technology, Multimedia Technology, Ericsson Research, Corporate Unit, Lulea, Sweden.
3. Atal, Bishnu S., and Suzanne L. Hanauer. 1971. Speech analysis and synthesis by linear prediction of the speech wave. *The Journal of the Acoustical Society of America* 50 (2B): 637–655.
4. Atal, Bishnu S., and Manfred R. Schroeder. 1970. Adaptive predictive coding of speech signals. *The Bell System Technical Journal* 49 (8): 1973–1986.
5. Atal, Bishnu S., and Manfred R. Schroeder. 1979. Predictive coding of speech signals and subjective error criteria. *IEEE Transactions on Acoustics, Speech and Signal Processing* 27 (3): 247–254.
6. Barbu, Tudor. 2004. Discrete speech recognition using a Hausdorff-based metric. In Proceedings of the 1st International Conference of E-Business and Telecommunication Networks, ICETE, Setubal, Portugal, 363–368.
7. Batliner, Anton, Richard Huber, Heinrich Niemann, Elmar Nöth, Jörg Spilker, and Kerstin Fischer. 2000. The recognition of emotion. In *Verbmobil: Foundations of speech-to-speech translation,* ed. Wolfgang Wahlster, 122–130. Berlin: Springer.
8. Burges, Christopher J. C.. 1998. A tutorial on support vector machines for pattern recognition. *Data Mining and Knowledge Discovery* 2 (2): 121–167.
9. Burkhardt, Felix, and Walter F. Sendlmeier. 2000. Verification of acoustical correlates of emotional speech using formant-synthesis. Paper presented at ISCA Tutorial and Research Workshop (ITRW) on Speech and Emotion, Newcastle.
10. Burkhardt, Felix, Astrid Paeschke, Miriam Rolfes, Walter F. Sendlmeier, and Benjamin Weiss. 2005. A database of German emotional speech. *Interspeech* 5:1517–1520.
11. Cowie, R., E. Douglas-Cowie, B. Apolloni, J. Taylor, A. Romano, and W. Fellenz. 1999. What a neural net needs to know about emotion words. In *Computational intelligence and applications,* ed. N. Mastorakis, 109–114. Wisconsin: WSEAS Press.
12. Cowie, Roddy, Ellen Douglas-Cowie, Nicolas Tsapatsoulis, George Votsis, Stefanos Kollias, Winfried Fellenz, and John G. Taylor. 2001. Emotion recognition in human-computer interaction. *IEEE Signal Processing Magazine* 18 (1): 32–80.
13. Cristianini, Nello, and John Shawe-Taylor. 2000. *An introduction to support vector machines and other kernel-based learning methods.* Cambridge: Cambridge University Press.
14. Charles Darwin. 1998. *The expression of the emotions in man and animals.* Oxford: Oxford University Press.
15. Douglas-Cowie, Ellen, Roddy Cowie, Ian Sneddon, Cate Cox, Orla Lowry, Margaret Mcrorie, Jean-Claude Martin, Laurence Devillers, Sarkis Abrilian, Anton Batliner, et al. 2007. The humaine database: Addressing the collection and annotation of naturalistic and induced emotional

© The Author(s) - SpringerBriefs 2015
K. R. Anne et al., *Acoustic Modeling for Emotion Recognition,*
SpringerBriefs in Electrical and Computer Engineering, DOI 10.1007/978-3-319-15530-2

data. In *Affective computing and intelligent interaction,* eds. Ana C. R. Paiva, Rui Prada, and Rosalind W. Picard, 488–500. Berlin: Springer.

16. Dubuisson, M.-P., and Anil K. Jain. 1994. A modified Hausdorff distance for object matching. In Proceedings of the 12th IAPR International Conference on Pattern Recognition, 1994. Vol. 1-Conference A: Computer Vision & Image Processing 1:566–568. IEEE.

17. El Ayadi, Moataz, Mohamed S. Kamel, and Fakhri Karray. 2011. Survey on speech emotion recognition: Features, classification schemes, and databases. *Pattern Recognition* 44 (3): 572–587.

18. Erickson, Donna. 2005. Expressive speech: Production, perception and application to speech synthesis. *Acoustical Science and Technology* 26 (4): 317–325.

19. France, Daniel Joseph, Richard G. Shiavi, Stephen Silverman, Marilyn Silverman, and D. Mitchell Wilkes. 2000. Acoustical properties of speech as indicators of depression and suicidal risk. *IEEE Transactions on Biomedical Engineering* 47 (7): 829–837.

20. Furui, Sadaoki. 2000. *Digital speech processing: Synthesis, and recognition.* Boca Raton: CRC.

21. Hansen, John H. L., and Douglas A. Cairns. 1995. Icarus: Source generator based real-time recognition of speech in noisy stressful and Lombard effect environments. *Speech Communication* 16 (4): 391–422.

22. Hozjan, Vladimir, Zdravko Kacic, Asuncion Moreno, Antonio Bonafonte, and Albino Nogueiras. 2002. Interface databases: Design and collection of a multilingual emotional speech database. 3rd International Conference on Language Resources and Evaluation, Las Palmas.

23. Huang, Cheng-Lung, and Chieh-Jen Wang. 2006. A GA-based feature selection and parameters optimization for support vector machines. *Expert Systems with Applications* 31 (2): 231–240.

24. Huang, Chun-Fang, and Masato Akagi. 2008. A three-layered model for expressive speech perception. *Speech Communication* 50 (10): 810–828.

25. Huang, Xuedong, Alex Acero, Hsiao-Wuen Hon, and Raj Reddy. 2001. *Spoken language processing: A guide to theory, algorithm, and system development.* Englewood Cliffs: Prentice Hall PTR.

26. Jain, Anil K., Robert P. W. Duin, and Jianchang Mao. 2000. Statistical pattern recognition: A review. *IEEE Transactions on Pattern Analysis and Machine Intelligence* 22 (1): 4–37.

27. Jesorsky, Oliver, Klaus J. Kirchberg, and Robert W. Frischholz. 2001. Robust face detection using the Hausdorff distance. In *Audio-and video-based biometric person authentication,* eds. Josef Bigün, Gérard Chollet, and Gunilla Borgefors, 90–95. Berlin: Springer.

28. Ji, Shuiwang, and Jieping Ye. 2008. Generalized linear discriminant analysis: A unified framework and efficient model selection. *IEEE Transactions on Neural Networks* 19 (10): 1768–1782.

29. Jurafsky, Daniel, and James H. Martin. 2000. *Speech and language processing: An introduction to natural language processing, computational linguistics, and speech recognition.* Englewood Cliffs: Prentice Hall.

30. Kecman, Vojislav. 2001. *Learning and soft computing: Support vector machines, neural networks, and fuzzy logic models.* Cambridge: MIT Press.

31. Kobayashi, Hajime, and Tetsuya Shimamura. 2000. A weighted autocorrelation method for pitch extraction of noisy speech. Proceedings of the 2000 IEEE International Conference on Acoustics, Speech, and Signal Processing, 2000. (ICASSP'00) 3:1307–1310. IEEE.

32. Koolagudi, Shashidhar G., and K. Sreenivasa Rao. 2012. Emotion recognition from speech: A review. *International Journal of Speech Technology* 15 (2): 99–117.

33. Kpalma, Kidiyo, Joseph Ronsin, et al. 2007 An overview of advances of pattern recognition systems in computer vision. In *Vision systems: Segmentation and pattern recognition,* eds. Goro Obinata and Ashish Dutta, 169–194. Rijeka: Intech.

34. Kuc, Roman. 2008. *Introduction to digital signal processing.* Hyderabad: BS Publications.

35. Kuchibhotla, Swarna, H. D. Vankayalapati, R. S. Vaddi, and K. R. Anne. 2014. A comparative analysis of classifiers in emotion recognition through acoustic features. *International Journal of Speech Technology* 17 (4): 401–408.

36. Liu, Chao-Chun, Dao-Qing Dai, and Hong Yan. 2007. Local discriminant wavelet packet coordinates for face recognition. *The Journal of Machine Learning Research* 8: 1165–1195.
37. Luengo, Iker, Eva Navas, Inmaculada Hernáez, and Jon Sánchez. 2005. Automatic emotion recognition using prosodic parameters. INTERSPEECH, Lisbon, 493–496.
38. Luengo, Iker, Eva Navas, and Inmaculada Hernáez. 2010. Feature analysis and evaluation for automatic emotion identification in speech. *IEEE Transactions on Multimedia* 12 (6): 490–501.
39. Ma, Jianhua. 2006. Ubiquitous intelligence and computing: Third International Conference, UIC 2006, Wuhan, China, September 3–6, 2006: Proceedings, vol. 4159. Springer.
40. Markel, J. 1972. Digital inverse filtering-a new tool for formant trajectory estimation. *IEEE Transactions on Audio and Electroacoustics* 20 (2): 129–137.
41. Menezes, Paulo, José Carlos Barreto, and Jorge Dias. 2004. Face tracking based on haar-like features and eigenfaces. IFAC/EURON Symposium on Intelligent Autonomous Vehicles, Lisbon.
42. Milton, A., S. Sharmy Roy, and S. Selvi. 2013. SVM scheme for speech emotion recognition using MFCC feature. *International Journal of Computer Applications* 69 (9): 35–39.
43. Murray, Iain R., and John L. Arnott. 1993. Toward the simulation of emotion in synthetic speech: A review of the literature on human vocal emotion. *The Journal of the Acoustical Society of America* 93:1097.
44. Nwe, Tin Lay, Say Wei Foo, and Liyanage C. De Silva. 2003. Speech emotion recognition using hidden Markov models. *Speech Communication* 41 (4): 603–623.
45. Orfanidis, Sophocles J. 1985. *Optimum signal processing: An introduction*. Collier Macmillan: New York.
46. Paleari, Marco, and Christine L. Lisetti. 2006. Toward multimodal fusion of affective cues. Proceedings of the 1st ACM international workshop on human-centered multimedia, 99–108. ACM.
47. Paleari, Marco, Ryad Chellali, and Benoit Huet. 2010. Features for multimodal emotion recognition: An extensive study. 2010 IEEE Conference on Cybernetics and Intelligent Systems (CIS), 90–95. IEEE.
48. Parasher, Mayank, Shruti Sharma, A. K. Sharma, and J. P. Gupta. 2011. Anatomy on pattern recognition. *Indian Journal of Computer Science and Engineering (IJCSE)* 2 (3): 371–378.
49. Perlibakas, Vytautas. 2004. Distance measures for PCA-based face recognition. *Pattern Recognition Letters* 25 (6): 711–724.
50. Poh, Norman, Thirimahos Bourlai, and Josef Kittler. 2009. Multimodal information fusion. In *Multimodal signal processing: Theory and applications for human-computer interaction*, eds. Jean-Philippe Thiran, Ferran Marqués, and Hervé Bourlard, 153–169. Amsterdam: Elsevier.
51. Polikar, R. 2006. Pattern recognition. In *Wiley encyclopedia of biomedical engineering*, ed. M. Akay. New York: Wiley.
52. Press, William H. 2007. *Numerical recipes 3rd edition: The art of scientific computing*. Cambridge: Cambridge University Press.
53. Rabiner, Lawrence R., and Ronald W. Schafer. 1978. *Digital processing of speech signals*. Vol. 100. Englewood Cliffs: Prentice Hall.
54. Ravikumar, M., and M. Suresha. 2013. Dimensionality reduction and classification of color features data using SVM and KNN. *International Journal of Image Processing* 1 (4): 16–21.
55. Ross, Arun, and Rohin Govindarajan. 2004. Feature level fusion in biometric systems. Proceedings of Biometric Consortium Conference (BCC).
56. Ross, Arun A., Karthik Nandakumar, and Anil K Jain. 2006. *Handbook of multibiometrics*. Vol. 6. New York: Springer.
57. Sato, Nobuo, and Yasunari Obuchi. 2007. Emotion recognition using mel-frequency cepstral coefficients. *Information and Media Technologies* 2 (3): 835–848.
58. Schafer, Ronald W., and Lawrence R. Rabiner. 1970. System for automatic formant analysis of voiced speech. *The Journal of the Acoustical Society of America* 47 (2B): 634–648.
59. Scherer, Klaus R. 1979. Nonlinguistic vocal indicators of emotion and psychopathology. In *Emotions in personality and psychopathology*, ed. Carroll Izard, 493–529. New York: Springer.

60. Scherer, Klaus R. 2003. Vocal communication of emotion: A review of research paradigms. *Speech Communication* 40 (1): 227–256.
61. Scherer, Klaus R., Rainer Banse, Harald G Wallbott, and Thomas Goldbeck. 1991. Vocal cues in emotion encoding and decoding. *Motivation and Emotion* 15 (2): 123–148.
62. Schröder, Marc. 2004. *Speech and emotion research*. Saarbrücken: Universit at des Saarlandes.
63. Schuller, Björn, Gerhard Rigoll, and Manfred Lang. 2004. Speech emotion recognition combining acoustic features and linguistic information in a hybrid support vector machine-belief network architecture. Proceedings of the IEEE International Conference on Acoustics, Speech, and Signal Processing, 2004. (ICASSP'04). Vol. 1, I–577. IEEE.
64. Schuller, Björn, Stephan Reiter, and Gerhard Rigoll. 2006. Evolutionary feature generation in speech emotion recognition. 2006 IEEE International Conference on Multimedia and Expo. 5–8. IEEE.
65. Schuller, Björn, Gerhard Rigoll, Michael Grimm, Kristian Kroschel, Tobias Moosmayr, and Günther Ruske. 2007. Effects of in-car noise-conditions on the recognition of emotion within speech. *Fortschritte der Akustik* 33 (1): 305.
66. Sidorova, Julia. 2009. Speech emotion recognition with tgi+. 2 classifier. Proceedings of the 12th Conference of the European Chapter of the Association for Computational Linguistics: Student Research Workshop. 54–60. Association for Computational Linguistics.
67. Sim, Dong-Gyu, Oh-Kyu Kwon, and Rae-Hong Park. 1999. Object matching algorithms using robust hausdorff distance measures. *IEEE Transactions on Image Processing* 8 (3): 425–429.
68. Sundararajan, D. 2001. *The discrete Fourier transform: Theory, algorithms and applications*. Singapore: World Scientific.
69. Van Bezooijen, Renee. 1984. *Characteristics and recognizability of vocal expressions of emotion*. Vol. 5. Berlin: Walter de Gruyter.
70. Vankayalapati, H. D., K. R. Anne, and K. Kyamakya. 2010. Extraction of visual and acoustic features of the driver for monitoring driver ergonomics applied to extended driver assistance systems. In *Data and mobility,* eds. J. Düh, Hartwig Hufnagl, Erhard Juristsch, Reinhard Pfliegl, Helmut-Klaus Schimany, and Hans Schönegger, 83–94. Berlin: Springer.
71. Vapnik, V. N. 1982. *Estimation of dependences based on empirical data*. New York: Springer.
72. Veeramachaneni, Kalyan, Weizhong Yan, Kai Goebel, and Lisa Osadciw. 2007. Improving classifier fusion using particle swarm optimization. *IEEE Symposium on Computational Intelligence in Multicriteria Decision Making*, 128–135. IEEE.
73. Ververidis, Dimitrios, and Constantine Kotropoulos. 2003. A state of the art review on emotional speech databases. In Proceedings of 1st Richmedia Conference, 109–119.
74. Ververidis, Dimitrios, and Constantine Kotropoulos. 2006. Emotional speech recognition: Resources, features, and methods. *Speech Communication* 48 (9): 1162–1181.
75. Vogt, Thurid, Elisabeth André, and Johannes Wagner. 2008. Automatic recognition of emotions from speech: A review of the literature and recommendations for practical realisation. In *Affect and emotion in human-computer interaction,* eds. Peter Christian and Russell Beale, 75–91. Berlin: Springer.
76. Wayman, James, Anil Jain, Davide Maltoni, and Dario Maio. 2005. An introduction to biometric authentication systems. In *Biometric systems,* eds. Wayman, James, Anil Jain, Davide Maltoni, and Dario Maio, 1–20. London: Springer.
77. Ye, Jieping, Tao Xiong, Qi Li, Ravi Janardan, Jinbo Bi, Vladimir Cherkassky, and Chandra Kambhamettu. 2006. Efficient model selection for regularized linear discriminant analysis. Proceedings of the 15th ACM International Conference on Information and Knowledge Management, 532–539. ACM.
78. Yegnanarayana, B., and R. Veldhuis. 1998. Extraction of vocal-tract system characteristics from speech signals. *IEEE Transactions on Speech and Audio Processing* 6 (4): 313–327.